Reading

Technology and Skills in Zimbabwe's Manufacturing: From Autarky to Competition

Also by Simón Teitel

INDUSTRIAL AND TECHNOLOGICAL DEVELOPMENT

RESOURCES, INDUSTRIALIZATION AND EXPORTS IN LATIN AMERICA (*with E. Londero et al.*)

TECHNOLOGY AND ENTERPRISE DEVELOPMENT: Ghana under Structural Adjustment (*with S. Lall et al.*)

TOWARDS A NEW DEVELOPMENT STRATEGY FOR LATIN AMERICA

TRADE, STABILITY, TECHNOLOGY AND EQUITY IN LATIN AMERICA (*with M. Syrquin*)

Technology and Skills in Zimbabwe's Manufacturing

From Autarky to Competition

Simón Teitel

First published in Great Britain 2000 by
MACMILLAN PRESS LTD
Houndmills, Basingstoke, Hampshire RG21 6XS and London
Companies and representatives throughout the world

A catalogue record for this book is available from the British Library.

ISBN 0–333–65224–X

First published in the United States of America 2000 by
ST. MARTIN'S PRESS, INC.,
Scholarly and Reference Division,
175 Fifth Avenue, New York, N.Y. 10010

ISBN 0–312–23290–X

Library of Congress Cataloging-in-Publication Data
Teitel, Simón.
Technology and skills in Zimbabwe's manufacturing : from autarky to competition / Simón Teitel.
p. cm.
Includes bibliographical references and index.
ISBN 0–312–23290–X
1. Manufacturing industries—Zimbabwe. 2. Competition, International. I. Title.

HD9738.Z55 T45 2000
338.4'767'096891—dc21
99–088978

10 9 8 7 6 5 4 3 2 1
09 08 07 06 05 04 03 02 01 00

Printed and bound in Great Britain by
Antony Rowe Ltd, Chippenham, Wiltshire

Contents

PART V CONCLUSIONS AND POLICY INTERVENTIONS

Preface

This study[1] to assess patterns of technology and skills acquisition and operation in Zimbabwe's industry originated in a survey of manufacturing firms, part of a series of similar studies undertaken in various Sub-Saharan Africa countries.[2]

Given the high rates of population growth presently prevailing in these countries, very substantial rates of economic growth will be required to attain a significant improvement in their per capita income. Thus the contribution to economic growth of potentially leading sectors (i.e. those growing much faster than the average) needs to be expanded. Among such sectors, manufacturing industry has a high priority claim.

Thus these countries will not be able to just continue to rely mostly on their relatively abundant agricultural and mineral resources, plus ample supply of low-cost unskilled labour, to pursue economic growth.

However, African countries are being urged to liberalize and to open up further their economies to international trade at a time when international competition in industry is extremely keen. As a result, the kind of industrial development required, i.e. efficient and export-oriented, may confront severe obstacles in a country such as Zimbabwe.

In the first place, its industrialists had become accustomed to the protected industrial development that blossomed when the country was under an economic blockade. Second, strong nationalist and socialist tendencies were shown by its government after independence. Third, competitive industrial development demands continuous access to technology and technical skills not now present, or forthcoming shortly, to attain international levels of cost and quality.

One of the key objectives of the study described in this book was to elicit primary information on the technological status of Zimbabwe's manufacturing establishments and the stock of technical skills presently at their disposal.

Detailed information on the technological origin of the enterprises, their operational capabilities, R&D involvement, and the technical skills at their disposal, was sought by conducting plant visits and carefully planned interviews – with top and production management in a sample of firms belonging to four preselected manufacturing sectors. A separate set of visits and interviews were carried out at selected industrial support institutions.

It was expected that the resulting data, properly analyzed and compared with that of other countries and sources, would permit the assessment of the level of technical capabilities presently available in Zimbabwe's manufacturing firms, as well as their main technical problems and needs.

Due consideration was given in the study's design to potential explanatory factors of differential enterprise performance such as: size, nationality, ownership structure, and ethnic origin.

There was of course awareness of the limitations involved in working with a limited number of firms, distributed among four sectors, and also stratified according to size, location, ownership, and ethnic origin.

Nevertheless, it was expected that the qualitative richness of the information to be obtained and its appropriate analysis would make an important contribution to our knowledge, quite limited at the present time, about the actual operation of manufacturing enterprises in Zimbabwe; particularly, their technical strengths, problems, and prospects.

In turn, the information developed led to the formulation of conclusions and recommendations to guide potential policy interventions as well as the provision of technical assistance and other aid.

Chevy Chase, Maryland SIMÓN TEITEL

Acknowledgements

The cooperation received from many individuals is gratefully acknowledged. Francisco Thoumi collaborated in the interviews and in preparing the study. Moses Tekere and Takawira Mumvuma, both from the University of Zimbabwe, participated in the field work. The questionnaire used for the survey was a modified version of one prepared with the help of Ricardo Soifer that had been applied in a similar study conducted with his collaboration in Kenya. Suresh de Mel and Koffie Nassr provided excellent research assistance.

Comments on previous versions of the study made by Elio Londero, Francisco Thoumi and Hugh Schwartz, as well as those received during presentations at World Bank meetings in Washington DC and Paris, and at UNCTAD and UNIDO seminars, held in Geneva and Vienna respectively, are duly appreciated.

Special thanks are due to the following public and private industrial support institutions visited in Zimbabwe and to their staff: Standards Association of Zimbabwe (SAZ), Harare Polytechnic, Faculty of Engineering of the University of Zimbabwe, Zimbabwe Institute of Engineers, Small Enterprises Development Corporation (SEDCO), Scientific and Industrial Research and Development Centre (SIRDC), and Zimbabwe Investment Centre (ZIC).

The greatest debt is due to the firms that so generously gave of their time during our visits to their plants and other facilities. Unfortunately, the confidentiality promised to them precludes acknowledging their names.

S.T.

Part I

Country and Manufacturing Industry Background

1
Country Background

1.1 History

Zimbabwe's historical background, particularly its settler and Southern Rhodesian past, affects its present economic performance. Key features of its recent history of particular relevance to industrialization and technological development are examined below, distinguishing three main periods: (i) the colony, (ii) the interval following the 'unilateral declaration of independence' (UDI), and (iii) the stretch after the final declaration of independence.

Colonial legacy

Although Zimbabwe was never formally occupied as a colony, it experienced a similar process during the time its resources were commercially exploited via concessions granted by the British Crown to Cecil Rhodes and his companies. During what can be called the colonial, or pre-independence period ending in 1963, the economy was mainly geared to the exploitation of the country's natural resources for export and most investments and developments took place in mining and agriculture.[3]

While mining continues to be very important in the country's production and exports, the sector did not fulfil the high expectations of providing quick and fabulous riches, initially entertained by the early colonizers, and attention was thus turned to agriculture.

Infrastructure and institutional support were naturally also mainly oriented towards mining and agriculture. The domestic market for manufactures was small, and practically limited to the white community. Most blacks lived (and still do) in communal lands and, given their very low incomes, represented only a reduced market for manufactured products.

Few incentives existed for the development of manufacturing industries until the Second World War, when scarcities created the equivalent of protectionist policies. A number of manufacturing plants were installed at that time, generally by white immigrants or settlers' descendants, since the vast majority of local blacks had no access to the skills and the financing required.

Colonial society was stratified into four main racial groups: whites, blacks, coloured, and Asians. Non-blacks never accounted for more than 5% of the population, and coloured and Asians represent only about 0.5% of the total. More recently, the high income of Zimbabwe relative to some of its neighbouring countries has attracted black immigration. About 5% of the black population is considered to be of immigrant origin.

Although holding a clear majority, blacks do not constitute a cohesive group. Three-quarters of the black population is Shona and about one-sixth Ndebele.[4] These two tribal groups formed the basis for the two main liberation organizations, supported by China and the then USSR respectively, which after independence developed into the two main, recently merged, political parties, ZANU and ZAPU.

Little integration took place among the four main racial groups and Zimbabwe remained a society dominated politically, socially, and economically by the white minority who through customs and law, perpetuated arrangements in which race was the determining factor in most social interactions (Nelson, 1983, p. xxvii). Not surprisingly, white entrepreneurs relied on other whites when getting partners or hiring managerial personnel, in part because little or no alternatives existed, but also because it was easier for them to communicate with and trust 'their own'.

Unilateral Declaration of Independence (UDI)

Ian Smith's defiance of the United Kingdom and the international community by declaring independence in November of 1963, led to the imposition of a blockade on Rhodesia. However, South Africa (an important trading partner) and Portugal (that at the time still controlled Mozambique and Angola) permitted the country's access to their ports thus reducing to a substantial extent the effective power of the blockade.

But it could be argued that the blockade was really counteracted by a strong effort at import substitution industrialization (ISI). This attempt at autarky obviously had a number of ill-effects due to the allocation of resources to some inefficient activities, but it also led to the acquisition of troubleshooting and repair skills necessary to fix machines and reproduce spares that no longer could be imported. Thus substantial learning took place in a number of manufacturing industries, and, in not a few cases, the successful exporters of today acquired experience first supplying the domestic market under high-level protection.[5]

During UDI the government established strict market controls and regulations similar to those present in centrally-planned or war economies. Foreign exchange was controlled, and implicitly overvalued, while competition was sacrificed to avoid duplication. On the positive side, domestic investment as a share of GDP increased from 16% in 1965 to 29% in 1975.

The manufacturing sector grew during this time at an yearly average rate of 8%. In many ways, this experience was similar to that of Latin America during the periods of forced import substitution industrialization that took place at the time of the Great Depression and the Second World War.

The last years of UDI were particularly difficult. Mozambique had become independent, closing that route to the sea, and the oil price increases of the 1970s very adversely affected the country's terms of trade that fell 36% between 1975 and 1979. During these five years manufacturing output declined 14%.

Independence

This new phase, that started in 1980, was characterized by emphasis in indigenous development supported by heavy state intervention. The new government had a strong socialist bent, and it inherited not an open market economy, but a highly regulated one. Given its initial objectives, the UDI regulations served the new regime government quite well, and most of them were kept in place.

Although development of the para-statal sector was promoted, the private sector continued to operate in an economic environment similar to that it was accustomed to. Again, in some respects it was useful to have had the experience of para-statal enterprises. Many are inefficient, but some, particularly those managed by the Reserve Bank (Central Bank) are not. They have also fulfilled the role of employing black professionals to an extent only matched by the local branches of some large multinational enterprises.

The first two years after independence saw a sharp, once and for all, increase in exports as the blockade was lifted, as well as very high GDP growth (11% and 13%) resulting from the relaxation of the foreign exchange constraint. However, this rapid economic growth also boosted imports drastically, and the increased role of the state led to a large expansion in fiscal expenditures and a macro-economic imbalance. The situation was aggravated by one of the worst droughts on record in 1983 when agricultural output declined by 6%.

The government applied a stabilization programme that included a general wage freeze, tax increases, tight monetary policies and a devaluation. The economy recovered in 1984 and 1985, and external balance was achieved, but the internal imbalance remained as the government budget deficit hovered around 10% of GDP, largely due to increased expenditures on education, health care, drought relief, and subsidies to para-statal companies.

The improved 1985 performance was reflected in an optimistic outlook for the First Five Year National Development Plan (1986–1990). However, in 1986 another severe drought caused a decline of 12% in agricultural output that led to lower aggregate demand, lower agricultural exports, and a tightening in the foreign exchange constraint. The government reacted with a series of stabilization measures including another wage freeze, further cutbacks in foreign exchange allocations, attempts to reduce input costs in agriculture, and a partial liberalization of foreign investment. Between 1988 and 1990, the economy recovered, and GDP grew at 4.6%.

By 1990 it was clear that the performance of the economy during the first decade of independence had not been satisfactory. Indeed, if one excludes the first two years of independence when income grew on the strength of the blockade elimination, since 1982, GDP had grown at only 2.2% per annum, 1% less than the 3.2% population growth rate. Similarly, the share of real gross capital formation in GDP had declined continuously from almost 20% in 1981 to a paltry 11% in 1988. These difficulties prompted the government to make a radical policy change.[6]

However, the reasons behind the policy changes may not have seemed obvious to a majority of the population since the country had not experienced hyper-inflation or a grave foreign debt crisis, and a substantial amount of blame for the income decline could be placed on the weather. Because of this, it is not clear how wide-spread is the public support to a major systemic change that goes drastically against the ideology espoused by the revolutionary elements that fought for independence and control the government since then.

Interestingly, the economic reform programme, launched in early 1991, was titled the Second Five Year National Development Plan (1991–1995). It attempted to liberalize markets and to decrease the role of the state. It contemplated the elimination of the foreign exchange allocation system, the liberalization of foreign investment rules, a sharp drop in the government deficit, the elimination of subsidies to para-statal companies and their monopoly rights, the elimination of price controls except for a few basic foodstuffs, the liberalization of the labour and transport markets, the elimination of direct monetary controls on interest rates, changes in the system of promotion and salary increases in the public sector to relate wages to performance, elimination of some local regulations that prevent the development of small enterprises, and the development of rural resettlement programmes, increases in agricultural services to communal areas, and changes in the land tax to discourage the holding of unutilized land.

These far-reaching reforms began with the reduction of protection and efforts at price stabilization, but were adversely affected by the worst drought of the century in 1991–92 which decimated cattle ranching and had many other deleterious effects on economic activity, particularly, in the countryside. In spite of these adverse conditions, the government continued implementing the programme.

It is likely that external changes such as those in China and the former Soviet Union, the two strongest supporters of the liberation war, and the political changes in South Africa, plus the support provided by the IMF and the World Bank, had convinced the government of the need to liberalize the economy and compete in world markets. However, the government has not succeeded in accelerating agrarian reform and promoting black entrepreneurship. Faced with internal economic and social unrest it has recently engaged in military intervention in neighbouring Congo and internal repression of the press, homosexuals, and political opposition groups.

Blacks are now realizing that political independence is not synonymous with economic and social fortune and are becoming increasingly aware of the need for institutional change, as well as of the various financial, technological, and other limitations confronting indigenous entrepreneurs.

1.2 Geography and natural resources

Zimbabwe is a relatively small and landlocked country in south central Africa. Its nearest sea access (Indian Ocean) is to the east, at Mozambique's port of Beira. Wholly within the tropics, most of the country consists of a high and middle level plateau lying between 900 and 1,700 metres above sea level. Thus it enjoys an exceptionally temperate climate. Geographical, or natural, borders, are the Zambezi river to the north and the Limpopo river to the south and west. The so-called Eastern Highlands create another border in the east.

Politically, it limits with Zambia in the north-west, Mozambique in the east and north-east, Botswana in the south-west, and South Africa in the south. The population is about 12 million, and the total area, of a little less than 400,000 square kilometres, is almost the size of the state of California in the United States.

Mineral resources

Zimbabwe is very well-endowed with mineral resources. While for various historical, institutional and other reasons, agriculture is better developed so far than mining, the existence of substantial mineral ores in the underground is not in doubt. A curious, connected fact, is that Zimbabwe has the world's highest per capita number of deaths caused by lightning, which seems to be attracted by the high concentration of minerals in its soil.

Zimbabwe produces more than forty minerals; it is the third largest producer of asbestos and chrome ore in the world, and the fifth or sixth of gold. Other important mineral products include coal, iron ore, nickel, and copper. Of lesser importance are diamonds, silver, cobalt, and tin; there also seems to be potential for the development of platinum mines.

In 1993, mining and quarrying contributed 7.1% to the value of GDP, and exports of mineral and metal products (gold, ferro-alloys, nickel, iron and steel products, and asbestos) accounted for 31.5% of the value of total exports (*EIU Country Profile 1994–95*, page 25, and reference table 26).

Agriculture

As can be seen from Table 1.1, about 50% of Zimbabwe's land area is covered by forests, which accounts for its abundant supply of wood for the furniture industry and other applications. Although only

Table 1.1 Land area and use, Zimbabwe and other selected countries (1987–89)

Country	*Land (000 Ha)*	*Cropland (%)*	*Pastures (%)*	*Forests (%)*	*Wild (%)*
Ghana	23002	11.74	21.74	35.69	0
Kenya	56969	4.25	66.88	4.18	19.0
Zimbabwe	**38667**	**7.23**	**12.56**	**49.89**	**0**
Africa	2964138	6.29	30.05	23.15	27.0
South America	1752926	8.08	27.24	51.10	21.0
Asia	2731228	16.64	25.42	19.73	13.0

Note: The proportions assigned may not always add up to 100 due to the classification method used.
Source: *World Resources, 1992–93*, A Report by The World Resources Institute, New York, Oxford University Press, table 17.

about 7% of its area is cropland, given its size and small population, the per-capita land endowment is quite large compared to other Sub-Saharan Africa countries (see Table 1.2).

Zimbabwe's agriculture was reputed to be among the most efficient in the world, and some still think that Zimbabwe should be the bread basket of Southern Africa. This optimism is somewhat tempered when account is taken of the deleterious effects severe periodic droughts have had on agricultural production in recent years.

Main commodities exported are: tobacco, sugar, cotton-seed, and tea. In 1989 they respectively represented 60.7%, 12.3%, 9.6% and 5% of total Sub-Saharan Africa exports (measured in metric tons).

Table 1.2 Land area and land use per capita, 1987–89, Zimbabwe and other selected countries (Ha per capita)

Country	*Land*	*Cropland*	*Pastures*	*Forests*
Ghana	1.53	0.18	0.33	0.55
Kenya	2.37	0.10	1.58	0.10
Zimbabwe	**3.98**	**0.29**	**0.50**	**1.99**
Africa	4.61	0.29	1.38	1.07
South America	5.92	0.48	1.61	3.02
Asia	0.88	0.14	0.22	0.17

Source: See Table 1.1.

Zimbabwe also produces grains, in particular, maize and wheat. In 1987, its share in the output of all Sub-Saharan Africa of these cereals were 13.0% for wheat and 7.0% for maize. Horticulture has become very dynamic recently and Zimbabwe is now the third largest world exporter of roses.

The country also has good cattle ranching facilities and it exports small amounts of hides and meat, with the bulk being consumed locally. However, the recent liberalization of beef prices may result in larger export surpluses as domestic consumption of beef tends to fall and is being substituted, in part, by increased consumption of fish (imported frozen from Namibia), poultry, and pork.

The cattle herd has been estimated at around 6 million heads with about two-thirds located in communal areas. Many may have perished during the 1992 drought. Zimbabwe has a quota of about 9,000 tons/year of beef exports with the European Union which it normally expects to meet.

One of the key problems of the agricultural sector is the present concentration of land ownership. The government has promised land redistribution and has been buying land from large commercial farmers to subdivide for new settlements. Of course, besides the problem of a reluctant (largely white) group of owners generally unwilling to give up its holdings, there is also the question of the potential trade-offs between efficient exploitation with economies of scale, and the danger of transforming successful large efficient farms into small and inefficient exploitations.

In part this debate is academic since many large farms are not being exploited by their owners and they are the natural first targets of resettlement programmes by the government. According to the Land Acquisition Act, the government is allowed to take over any land for farm settlement and to decide compensation. Landowners are prevented from trying to reach an alternative settlement by legal means.[7]

1.3 Policy and market environment

Entrepreneurial performance in Zimbabwe is necessarily conditioned by the policy framework in which the market system operates. In addition to the government's economic policies and regulations, other existing social, political, and institutional constraints must also be included. Selected features, including existing market imperfections,

Table 1.3 Zimbabwe: economic indicators, 1992–97

Econ. indic.	*1992*	*1993*	*1994*	*1995*	*1996*	*1997*
Real GDP growth %	–5.5	2.0	5.3	–0.2	7.2	3.7[a]
Vol. of manuf. '80 = 100	129.9	119.3	130.7	112.9	116.9	114.9[b]
Ext. debt/GDP %	78.5	80.7	82.1	80.5	69.2	64.9[a]
Ex. rate (av) Z$:US$	5.10	6.48	8.15	8.66	9.92	11.89

Notes: [a] estimate
[b] actual value for June
Source: *EIU Country Report, Zimbabwe, 4th quarter 1998*, The Economist Intelligence Unit Limited, 1998.

racial attitudes, market orientation, and some political and other non-economic constraints, are examined below.

Adjustment

In 1990–91, following the advice of donors and international agencies, the government introduced an economic adjustment programme upon which future economic aid had been conditioned. Its strictures were similar to those adopted in other African countries and included: trade liberalization, privatization, reduction of the fiscal deficit, devaluation of the currency, promotion of foreign investment, reduction or elimination of price controls and other internal regulations. Price controls and wage regulations were ended in 1993. Exchange devaluations took place since the plan's inception, and by 1997, while moving towards convertibility, the local currency had plummeted to less than half its 1991 value. In 1994 most marketing boards were deregulated and a privatization programme was started.

As can be seen from Table 1.3, results in terms of economic growth have been uneven. Real GDP growth was –5.5% in 1992 as a result of the drought and recovered mildly to 2.0% in 1993. While a strong recovery took place in 1994 with a 5.3% growth in GDP, the ensuing recession resulted in a negative 0.2% rate for 1995 followed by a strong recovery to 7.2% in 1996, and an estimated 3.7% growth for 1997.

As can also be seen, the volume of manufacturing production continued to decline, particularly in recent years, and the fall in the value of the country's currency *vis-à-vis* the dollar has been quite dramatic (Figures 1.1 and 1.2). Of course, these results have been affected by the

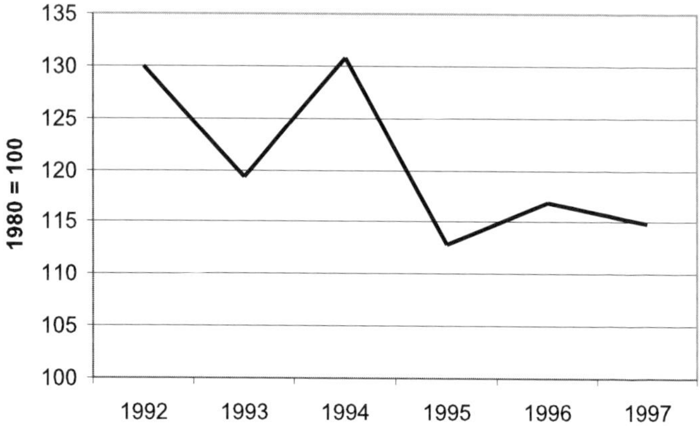

Figure 1.1 Zimbabwe. Index of manufacturing output, 1992–1997 (1980 = 100)

adverse effects of periodic severe droughts, plus the contraction due to the adjustment programme.

In 1995 the IMF suspended its support to the programme because the government did not meet its spending targets, and by 1996 had not endorsed yet the second phase of the reform plan put forward by Zimbabwe. The fiscal deficit remained above 10% of GDP in spite of real cuts in health, education and defence expenditures. Price inflation was still around 20%, and interest rates remained high. About two-thirds of the population was estimated to live in poverty during

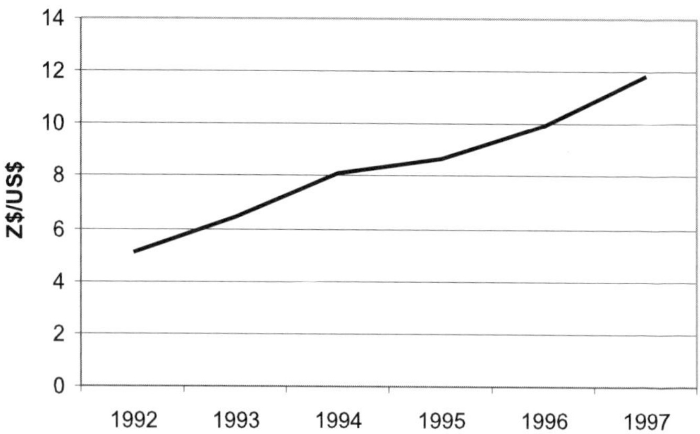

Figure 1.2 Zimbabwe. Average nominal exchange rate, 1992–1997 (yearly average in units of Zimbabwe's currency per US dollar)

1996, and there has been an increase in the incidence of malaria and AIDS is ravaging the adult black population.

The move towards a free market economy in Zimbabwe has been equated with that of socialist countries in East and Central Europe and in Asia (Cook and Nixson, 1995). It could also be seen as paralleling the recent experience of some Latin American countries following periods of trade protection and populist governments.

Market imperfections

The prevalence of economies of scale and indivisibilities in modern manufacturing, plus the country's limited market size for most industrial products, have resulted in a small number of producers for many goods which leads to imperfect competition. Geographical, social, political, and legal factors, such as location, social status, political affiliation, and other characteristics of market participants, as well as the legal and unwritten social rules regulating commercial transactions, are also responsible for important market imperfections.

Markets in Zimbabwe tend to be segmented across racial, tribal, political, ideological, and physical boundaries. These factors have allowed economic growth up to a point, but as the economy turns increasingly more diversified, they become an obstacle to further growth.

Social factors contribute to lack of trust across groups which in turn tends to increase transaction costs significantly. Social, political, racial, tribal, and other sources of market fragmentation allow high levels of trust within particular segments, but very little trust across the board. Thus, in attempting to reduce transaction and business costs, entrepreneurs often prefer to hire people belonging to their own group which they can trust more than members of other groups.

Racial attitudes

Quite different mentalities have developed within the Black and White communities in Zimbabwe. Many whites, particularly those born in the country who consider themselves Zimbabweans and do not see emigration as a realistic alternative, feel threatened and isolated from the world. On the other hand, many blacks feel that they have been deprived of their just share of income and wealth, and expect special treatment from the government.

After independence, many whites, particularly those with some skills, emigrated, but others, mainly those who had some property, remained, or emigrated and returned, as their Zimbabwean roots were very strong. Emigration for this group would entail a substantial drop in their standard of living and status, and severe psychological losses. It is very likely that the overwhelming majority of whites presently residing in Zimbabwe are there for the long run.[8]

On the other side, there is a black group that questions the roots of the whites' wealth and income. This group has developed an entitlement mentality that encourages rent-seeking and is not very conducive to economic growth.[9]

The trade embargo during UDI, and the foreign exchange policies followed after independence, have also contributed to the whites' feeling of isolation. Since UDI, and through the 1980s, the country faced a sharp foreign exchange constraint that affected economic behaviour in a number of ways. Firms that were able to buy imported spare parts or intermediate goods for locally produced goods with a high import content, tended to hoard them to minimize risks; firms that generated foreign exchange, tried to lay claim over it, and succeeded, at least partially, in having the government enact systems that gave them special preferences and access to foreign exchange.

Market orientation

Historically, as noted above, Zimbabwe's development has had a strong inward market orientation that began to change only in the last few years. Zimbabwe's import substitution strategy differed in its origins from those followed by most developing countries in that it was imposed, for political reasons, by the country's major trading partners.

Based in abundant natural resources, that by African standards provided a relatively high income per capita, and a cadre of relatively skilled people, even before independence the country had developed a comparatively more advanced manufacturing base than its neighbouring poorer countries. However, investment during this period was very limited. After independence, inward-looking policies continued, this time supported by the government's ideology, but the threat of nationalization and violence also discouraged investment.

Since 1965, when UDI was declared, until the 1991 economic policy changes, the main problem for manufacturing was to find ways to keep producing. Domestic market competition was weak and quality was not a key concern.[10] Still, some plants began to sell abroad, as the availability of natural resources, and the low cost of labour, provided large competitive advantages that allowed them to develop exports in spite of an overall anti-export policy bias.

Some of the policy changes that began to liberalize the economy in 1991 became necessary as the post-independence policies produced unsustainable macroeconomic disequilibria. However, as is the case in a number of other countries, these changes are perceived by some groups as imposed from abroad, and perhaps unnecessary, and are being implemented by a government with a political philosophy often in contradiction with the changes. Thus, policy implementation may at times appear to lack conviction which gives raise to credibility misgivings.

In addition to the local policy changes, other sources of uncertainty, but also of increased opportunities, are now confronting Zimbabwe's entrepreneurs. The recent political changes in South Africa not only open up a relatively large and potentially wealthy market just to the south of Zimbabwe, but also underline the presence of a strong competitor in domestic and some export markets. Moreover, the establishment of trade agreements with South Africa, and the formation of a Southern African Free Trade Association (SAFTA), are also perceived as developments that are bound to challenge Zimbabwe's industrial establishment.

Political and other non-economic constraints

While differences in treatment between black and white citizens are the most obvious source of social stratification in Zimbabwe, a number of other cleavages are also important, particularly those among black ethnic groups which have led to substantial violence in the past, and have left a legacy of mutual distrust among groups. Not only was the war for independence very bloody,[11] but the level of violence remained high after independence, as the fight for the government spoils spanned ethnic conflict.

Violence has subsided since the mid-1980s, and the country appears to be pacified. However, given the source of the violence, there is no guarantee that it will not surge again in the future. Furthermore, the two main ethno-linguistic groups, the Shona and the Ndebele, are also divided in subgroups that could become an additional source of conflict. In the long run, the extent to which ethnicity serves as a basis for division between Zimbabwe's two most important African peoples remains an important issue (Kaplan, 1983, p. 92).

The country also has strong economic ideology cleavages. On the one hand, political parties and government have been controlled by blacks and have espoused leftist rhetoric supported by various socialist ideologies. This is quite understandable, as the colonial experience contributed to the identification of capitalism with colonial power, and of socialism with independence. This perception was reinforced by the support that the USSR and China provided to the two main independence-seeking groups during the war for independence, and, perhaps, by some perceived similarities between socialism and traditional black community institutions.

On the other hand, the productive activity in the economy is largely controlled by whites, strongly committed to capitalism. Blacks talk socialism and control the polity, but whites talk capitalism and control the economy. This has produced a substantial divergence and conflict between acceptable political and business behaviours. While the political rhetoric has leftist overtones, most economic activity still takes place within a fairly capitalist framework.

The rhetoric and functioning of democracy also conflict with an economic reality in which transactions and policy decisions are often perceived by some groups as being arbitrary and influenced by social, political, and other non-economic factors. Both black and white societies have a tradition of inequality and authoritarian behaviour that may help to explain the extent of acquiescence to such practices observed in Zimbabwe.

The political party structure also influences the economic environment. Zimbabwe's authoritarian tradition is reflected in a low degree of tolerance for opposite views, and great difficulties to use compromise to build a political consensus. In the late 1980s, after a bloody conflict, the two prevailing parties agreed to form one, that controls the government. The dominant party is a peculiarly Zimbabwean institution that also engages in productive investments.

There is no question that the symbiotic relationship that exists between the ruling party and the government affects negatively the country's economic environment. While this relationship does not necessarily imply that the government is corrupt, since the system does not have strong checks and balances, clientelism and nepotism loom large and could easily become the rule.

According to some observers corruption has become a normal cost of business in Zimbabwe, and it is claimed that most government contracts now require some type of 'commission' for those controlling the regulations or possessing the necessary political influence to win approval. As ZANU PF has increased its electoral grip to the point of controlling 147 out of 150 seats in parliament, the distinctions between the state, the government, and the party have become quite blurred.[12] Strong allegations of business interests in the Congo Republic by members of the government involved in providing military support to the Congolese government, have also been raised.

2
Manufacturing Industry

2.1 Industrialization

Although Zimbabwe is a relatively small and poor (about US$650 of income per capita) developing country, compared to other Sub-Saharan Africa countries it has a relatively advanced manufacturing sector which in 1990 accounted for about one-quarter of GDP at factor cost – over twice the share of countries like Kenya and Ghana.

The relatively large size of Zimbabwe's manufacturing sector could be explained as resulting from both the country's factor endowment and its history. The large commercial agriculture sector created by white settlers during the first 65 years of this century led to the development of complementary agro-industries, and generated a market (albeit not large) for such manufactured products as tractors, diesel engines and fertilizers.[13]

After UDI in 1965, the external blockade provided a very strong incentive for the development of all sorts of manufactures. From 1964 to independence in 1980, manufacturing production volume grew at a compounded 5.5% annual rate[14] and product diversification increased dramatically. It has been estimated that by the time of independence, the manufacturing sector produced 7,000 different products, compared with 1,000 products in 1966 (Riddell, 1988).

The relatively higher development of Zimbabwe's manufacturing sector among Sub-Saharan Africa countries is also reflected in the structure of its merchandise exports. In spite of a very large primary factor endowment of minerals and good agricultural land, about one-third of exports are manufactures. Notably, in 1991 textiles and clothing accounted for 6% and machinery and transport equipment for 4% of total exports.

The fast product diversification and high output volume growth indicate, although no direct data is available, that during UDI manufacturing value added grew at a significantly higher rate than GDP, and import substitution advanced significantly. It was estimated that in 1969 manufacturing had become the largest economic sector contributing 19% of GDP. By 1980, this share reached 24.9%. During UDI, textiles and metal and metal products had the fastest production volume growth (7.5% and 7.3% respectively).[15] At the other end, transport equipment grew at only 1%, and clothing and footwear at 3.1% (Nelson, 1983, appendix).

The manufacturing growth pattern changed significantly after independence. GDP growth slowed down to 2.7% for the 1980–1989 period, while manufacturing grew at the same rate of GDP, so that, as indicated above, its GDP share in 1990 was equal to that of 1980. During this decade, sub-sectoral growth varied significantly from that during UDI. Textiles, non-metallic minerals, and chemicals and petroleum were the faster growing sub-sectors. Transport equipment grew very fast during the first two years, reflecting the repressed demand build up during UDI, collapsing afterwards and recovering also very fast in 1989. Wood and furniture was depressed, and metal and metal products and beverages and tobacco were stagnant during the 1980s. Foodstuffs, clothing and footwear, and paper and printing, grew at close to the average of the sector and the economy. The growth pattern differences reflect the elimination of the blockade that led to some export substitution, the larger role of the para-statal companies, the slow growth of the economy, and the reluctance of the private sector to invest.

The policy changes initiated in 1991 are likely to modify the composition of manufacturing. Although the time elapsed since the new policy measures were taken might be insufficient to detect major trends, sharp drops have already taken place in the volume of production of traditional import substitution sub-sectors (beverages and tobacco, textiles, clothing and footwear, chemical and petroleum products, and metal and metal products), and important increases in wood and furniture (whose exports have grown significantly), and paper, printing and publishing.[16]

Economic policies, and the long period of foreign exchange constraint experienced during UDI and the first decade of independence, led to a high market concentration in manufacturing. It was estimated in 1985 that 50% of the manufacturing products were produced by monopolies and 40% by oligopolies (World Bank, 1993b, based on UNIDO's data). There are indications that the recent policy changes have increased significantly the degree of competition faced by manufacturers in the country.

As would be expected, manufacturing is concentrated in the main urban centres. About half the manufacturing establishments are located in Harare and another quarter in Bulawayo. In 1990, Harare accounted for almost 53% of gross output of manufacturing and Bulawayo for about 20% of the total (EIU, 1995, p. 16). Manufacturing accounts for over 16% of formal sector employment, virtually all of which is urban. Since the two main cities are home to the largest plants, which tend to be more capital intensive and to enjoy economies of scale, they account for a smaller proportion of total manufacturing employment – 28% and 12.4% respectively in 1992 (EIU, ibid).

Nevertheless, manufacturing is a greater urban employer in Zimbabwe than in other developing countries, with similar and higher urban population shares and where the urban tertiary sector has absorbed most rural migrants. The importance of manufacturing activities in the urban sector also highlights the potentially huge effect that changes in manufacturing policies are bound to have on urban development in Zimbabwe.

Other than the wholesale incorporation of blacks to its entrepreneurial and managerial strata, the fundamental problem faced by Zimbabwe's manufacturing sector is how to redirect the creative abilities of its industrialists, that were quite successful in developing ways to work under highly-protected but constrained circumstances, to enable them to succeed in a much more open and competitive environment.

2.2 Factors of production

Raw materials

From the point of view of their potential use in manufacturing, the country could rely only on a few locally available raw materials. These include some agricultural resources as inputs to the food industry; cotton, for the textiles and garment industry, and forest products, specially pine and some hardwoods, for the wood and furniture industry. As for mining products, little finds its way into local manufacturing production. A local steel mill (ZISCO) produces some of the iron and steel basic intermediates for metalworking, but many other raw materials and intermediate products must be imported.

Consequently, during periods of restricted availability of foreign exchange, industrial production suffered severely. The same was true during the periodic droughts which ruin crops, decimate livestock, and have other indirect detrimental effects. The restricted size of domestic markets and the consequent lack of flexibility in sourcing, clearly accentuate the extent of dependency of manufacturing firms in Zimbabwe on their imported raw materials suppliers.

Thus, with the possible exceptions of the food and furniture industries, domestic natural resources do not play an important role as inputs in Zimbabwe's manufacturing industry.

Labour

There is of course a plentiful supply of unskilled labour in Zimbabwe. With an average rate of population growth of 2.9% during the 1970s, the consequent growth of the labour force has also been quite high and led to substantial open and disguised unemployment. Moreover, the increase in the rate of population growth to 3.4% that took place during the 1980s sends a foreboding message to economic planners and policy makers.

Nevertheless, Zimbabwe (together with Botswana) seems to have been in the vanguard of the demographic transition to lower population growth among countries in Sub-Saharan Africa. The decline in the demographic expansion has been linked to the higher educational levels attained by women in these countries (Thomas and Murandi, 1994).[17]

Education has increased tremendously in the last 20 years, particularly at the secondary level, where with an enrolment at present of about 50% of the population of the corresponding age, Zimbabwe has attained similar average levels to those prevailing in East Asia and Latin America. (See Table 2.1.)

In spite of the relatively high general education level of the population, and of being a resource-rich country, unskilled labour is still comparatively inexpensive in Zimbabwe thanks to the high rate of demographic growth. On the other hand, skilled labour is rather scarce and unevenly distributed. High-level skilled personnel, in particular engineers, scientists and well-trained technicians, are quite rare in Zimbabwe's manufacturing industries.

Government legislated job security seems to affect the demand for labour, specially in the organized, or formal, sector, and particularly among large firms that are expected to more closely obey government regulations (Fallon and Lucas, 1993).

Table 2.1 Zimbabwe: comparative educational attainment, 1970–90 (enrolment in percentages of target population)

Country	*Primary*		*Secondary*		*Tertiary*	
	1970	*1990*	*1970*	*1990*	*1970*	*1990*
Ghana	64	75	14	39	2	2
Kenya	58	94	9	23	1	2
Zimbabwe	**74**	**117**	**7**	**50**	**1**	**5**
SSA	46	68	6	17	1	2
Bolivia	76	82	24	34	17	23
Colombia	108	110	25	52	10	14
Peru	107	126	31	70	19	36
Latin America	95	107	28	49	15	16
Indonesia	80	117	16	45	n.a.	n.a.
Philippines	108	111	46	73	3	27
Sri Lanka	99	107	47	74	3	4
East Asia	88	127	24	49	4	5

Source: World Bank, *World Development Report 1993*, table 29, pp. 294–5.

Small enterprises seem to be responsible for a substantial proportion of the increase in employment, and most of the increase is due to start-ups of new enterprises with little growth originating in existing businesses (Mead, 1994).

Capital stock

Because prices and policy incentives under which manufacturing plants have operated for many years have been quite biased against investments in machinery and equipment, it was to be expected that Zimbabwe's manufacturing capital equipment be quite old. During UDI it was difficult to obtain new machinery, and in the post-independence period there has been little investment in fixed capital formation. Thus, it is not surprising to find machines from the pre-UDI period still in use.

During most of the post-independence period, besides the obvious risk of political instability and the threat of expropriation, private manufacturing has operated under heavy regulation of labour, goods, and capital markets. Private entrepreneurs had strong incentives to maximize the cash flow taken out of the plants while investing the minimum possible in plant maintenance and in new equipment. Furthermore, a large proportion of mid-level technical workers left the country.

On the other hand, para-statal companies faced a quite different set of incentives. Until 1991 they were able to import capital equipment duty free, and could easily bypass the foreign exchange restrictions that affected the private sector.

During the UDI period, the international blockade and political isolation made it difficult to produce, but it also made it necessary to carry out some investment in order to maintain output. The situation also promoted the development of mechanical skills to maintain old machines, make spares parts, and improvise ways to use machines to do tasks for which they were not originally conceived.

Excessive reliance on old-aged machinery and equipment, as occurred for a long time in Zimbabwe, may have both positive and negative effects. On the positive side, in such a circumstance capital costs tend to be minimal and manufacturing costs are practically limited to those of labour, intermediate goods, and services. Since labour costs are generally low, one can expect to find manufacturing plants with very old capital equipment and low labour productivity that can produce competitively – even for international markets, in certain lines of production.[18] Also, although the old machines pose maintenance skills demands, they tend to embody simpler manufacturing technologies that are an appropriate complement to the low-skills labour prevailing in the country.

On the negative side, old machines frequently limit product diversification and make it very difficult to produce high quality products. Some machines cannot perform as many tasks as newer models that use the same basic technology, and old machines also tend to lower the quality of some products. Because of normal wear and tear due to age, they also tend to produce a relatively high proportion of irregular, or slightly defective, products that do not meet the specifications of upscale markets.

As noted, the old machines and the previously existing foreign exchange constraints for importing new machines, have tested the ingenuity of Zimbabweans who learned to produce with very little capital. However, this state of affairs has also deterred product and process innovation, organization and management skills development, and, in general, the modernization of manufacturing.

Finally, the old equipment has also discouraged the adoption of safety and pollution control measures. The latter have generally been adopted in the developed world during the last decades, and were normally not incorporated into the designs of older machines.

Organization and management

Manufacturing in Zimbabwe presents a diversity of ownership, organizational and managerial structures. Most plants belong to (white) individual entrepreneurs; others, particularly the largest ones, are part of foreign corporations or group holdings, mostly from the United Kingdom and South Africa. Para-statal firms grew after independence, some firms were created as cooperatives, and as usual (and across ownership regime), firm size varies substantially.

In spite of the diversity of ownership systems, during UDI and independence all manufacturing plants have been subject to similar pressures and constraints, with the possible exception of differential access to foreign exchange which, as noted, favoured para-statal firms.

Managerial structures have, of course, been affected by historical developments. After independence, many mid-level white managers left the country, as did a large number of recently trained young white people. Since independence, the number of technical personnel graduating from the University of Zimbabwe and Harare Polytechnic has increased significantly. However, not many of them have been incorporated into the managerial structures of manufacturing firms.

A very important characteristic of the organizational and managerial structures of many firms in Zimbabwe is the scarcity of black managers.[19] There is no doubt that for the country's industry to be able to compete in a more open economy it will be necessary to develop black managerial skills, and to increase substantially the number of black managers. The country simply cannot afford to write off 95% of the population as a potential source for its managerial cadres. One of the main policy challenges faced by both the government and the private sector, is how to expand black participation in management without increasing state or party ownership.

2.3 Policy

While it has been argued that industrial development proceeded in Zimbabwe without the help of any direct incentives (Whiteside, 1989), there is no doubt that first under the blockade, and later on under UDI and independence, industry benefitted from a substantial measure of protection. Moreover, since independence, there has been a significant bias towards nationalism as well as government intervention and ownership, particularly in certain sectors. Although this is of course changing since the liberalization programme started in 1990–91, the government continues to propagate two rethorically very different policy messages; one, for domestic consumption, emphasizing indigenization and redistribution, and the other, for the foreign community, talking about privatization, government restraint, and liberalization of the economy.

The lack of incentives has also been blamed for the limited inflow of direct foreign investment, which is apparently being attracted by countries such as Botswana, that provides, in addition to stable economic policies, tax incentives and training grants, while also trying to minimize red tape and excessive regulations that continue to be prevalent in Zimbabwe.

Although protected for 25 years, the reaction to liberalization has been generally favourable among the large manufacturing firms, while smaller firms, particularly in those sectors more affected by import competition, have reacted very negatively (Skalnes, 1993). Clearly the adjustment cum liberalization programme has implied a shift towards less value added by local manufacturing, increased imports, and a greater relative economic contribution made by locally produced primary goods (Davies and Rattso, 1996). It has even been argued that there is a possible antinomy between macroeconomic policy reform, as advocated for Zimbabwe and other developing countries, and industrialization (Stoneman, 1992).

The economic recovery during 1996, with growth of 7.2% in GDP, was mostly centred in mining and agriculture. The textile industry continued to experience serious difficulties, and manufacturing output, as a whole, continued to be depressed.[20] In fact, during 1995 it was the lowest in a decade, with little or no recovery during 1996 and 1997 (Table 1.3).

3
Education, Science and Technology

Zimbabwe is not only better off than most other African countries as far as the educational attainment of its population, but also in terms of the stock of scientists and engineers, and various measurable outputs of scientific and technological activity. This very good relative performance provides little consolation however, since Africa lags substantially behind other continents and developing areas in these respects. At the same time, it constitutes an indication of potential capacity for future technological development in the country.

3.1 Education

Education attainment data provides a first indication of the country's potential for scientific and technological development. As previously shown in Table 2.1, in the last two decades, Zimbabwe has achieved substantial growth in enrolment at all educational levels. The enrolment expansion at the secondary and tertiary levels, in particular, has been quite dramatic (specially after independence) and the levels attained by the country far surpass the average for Sub-Saharan Africa.

Nevertheless, as can also be gathered from Table 2.1, in secondary education enrolment, Zimbabwe still lags behind such not overly developed countries as Peru in Latin America, and the Philippines and Sri Lanka in East Asia. The lag seems to be even more severe in tertiary education, given that, for example, relatively poor countries such as Bolivia in Latin America, and the Philippines in East Asia, held in 1990 much higher enrolment proportions.

3.2 Science and technology inputs

In 1985, Africa, excluding the Arab states, had one-tenth the number of scientists, engineers and technicians in Latin America and the Caribbean (469,000 versus 4.75 million: UNESCO, 1993, table 5-1). On a per capita basis, the numbers were approximately 1.38 and 11.76 per thousand population, respectively.

As to scientists and engineers engaged in R&D, Africa, again excluding the Arab states, had in 1990, 0.7% of the world total number of scientists and engineers engaged in R&D, while Latin America and the Caribbean had 3.1%. In terms of R&D expenditures, the share of Africa was 0.2% and that of Latin America and the Caribbean, 0.6% of the world total expenditures. As a proportion of GNP, Africa spent in R&D in 1990, 0.29% while Latin America and the Caribbean spent 0.40%.

No reliable data are available for Zimbabwe in this respect, but it is safe to assume that given its income per capita, which is above the average for South Saharan Africa countries, and its educational attainment in the secondary and tertiary field, the country should be above regional averages.

3.3 Science and technology outputs

While it is admittedly hard to quantitatively assess the results of scientific and technological activity, data on some indicators exist which permit at least rough comparisons with other countries and areas. Two such science and technology output indicators are: the number of patents granted and the number of scientific papers published in the international literature.[21]

Patents

Table 3.1 shows the number of patents granted, to residents and non residents, in Zimbabwe and other African countries, for the longest historic period with available information. Table 3.2 shows similar data, on a per capita basis, available for 1985 for Zimbabwe and other developing countries.

Several conclusions follow from these data. First, the output of industrial research, in terms of patents granted is quite low in Sub-Saharan Africa countries, including Zimbabwe.[22] Second, all countries included, except Kenya, have lower yearly average number of patents granted recently than during the historic period (generally from the

Table 3.1 Patents granted in Zimbabwe and other SSA countries (circa historic period 1960–82)

Country	*Historic period*	*Patents granted to residents*	*Patents granted to non residents*	*Patents granted total*	*Average yearly count*
Ghana	1963–82	0	1155	1155	57.7
Kenya	1965–82	6	1901	1907	105.9
Nigeria	1964–82	24	4765	4789	252.0
Tanzania	1955–82	0	1678	1678	59.9
Uganda	1963–82	1	1038	1039	51.9
Zambia	1966–82	24	2730	2754	162.0
Zimbabwe	**1958–82**	**368**	**9985**	**10353**	**414.1**

Source: World Intellectual Property Organization, *100 Years Protection of Industrial Property – Statistics – 1883–1982*, Geneva, 1983.

early 1960s to 1982). Third, the vast majority of the patents are granted to non-residents, thus not representing industrial research activity carried out in the country. Even those granted to residents could belong to subsidiaries of MNCs. Fourth, on a per capita basis, for total number of patents, Zimbabwe compares favourably with other, relatively poor, countries in Asia and Latin America, although the extremely low proportion of patents granted to residents should be noted.

Table 3.2 Comparison of patents granted in 1985, Zimbabwe and selected countries

Country	*Residents*	*Total*	*Proportion residents %*	*Population (millions)*	*Total pat. per mill.*	*Res. pat. per mill.*
Ghana	0	1	0	12.720	0	0
Kenya	0	98	0	20.330	4.82	0
Zimbabwe	**2**	**213**	**0.01**	**8.380**	**25.42**	**0.24**
Bolivia	3	62	4.80	6.430	9.64	0.47
Peru	29	155	18.70	19.420	7.98	1.49
Philippines	22	1281	1.70	54.670	23.43	0.40
Sri Lanka	18	112	16.10	15.840	7.07	1.14

Sources: Patents: World Intellectual Property Organization, *Industrial Property Statistics* 1985, Geneva, 1986. Population: Unesco, *op. cit.* table 1.1.

Table 3.3 Research publications of selected SSA countries, 1981–86

Country	*Total 81–86*	*Mean 81–86*	*% of total*	*Cumul. %*
Nigeria	4529	755	47	47
Kenya	1454	242	15	62
Zimbabwe	**407**	**68**	**4**	**66**
Tanzania	393	65	4	70
Ivory Coast	290	48	3	73
Zambia	172	29	2	75
Other	2400	400	25	100
Total	9645	1607	100	–

Source: Zymelman, 1990, table I-4.

Scientific papers

The production of scientific literature represents another potentially useful indicator of scientific and technological activity. Data on scientific papers published is, however, scarce for developing countries, and particularly so for African countries. Nevertheless, information on some African countries has been made available recently and is reflected in Tables 3.3 and 3.4.

The total Sub-Saharan Africa output of published scientific papers amounts, approximately, to 0.5% of world output. This compares unfavourably with its share of world population (9.1%) and GDP (0.76%). Latin America, which accounts for only 8.5% of world population but has a larger share of world GDP (5.6%) than Sub-Saharan Africa, is responsible for about 1.1% of world scientific output (Inter-American Development Bank, 1988, table IX-1), i.e. more than double Sub-Saharan Africa's share.

The data in Table 3.3 shows that African research publications are concentrated in a few countries. The two major producers of published scientific papers account for more than 60% of total Sub-Saharan Africa output, and six countries are responsible for three-quarters of the total. Zimbabwe's share of 4% is somewhat above its participation in both GDP (3.3%) and population (2.06%) in Sub-Saharan Africa.

Table 3.4 shows the distribution of research papers by field for selected Sub-Saharan Africa countries, including Zimbabwe, and also indicates how this distribution compares with those corresponding to Latin America and the world. As can be seen, the field distribution of research publications in Sub-Saharan Africa countries is quite

Table 3.4 Distribution of research publications by field; selected SSA countries, Latin America and the world (percentages)

Field	*Nig*	*Ken*	***Zim***	*Tan*	*IvC*	*Zam*	*Mean SSA*	*LA*	*World*	*AI SSA**	*AI Zim**
Clinical Medicine	38	61	**32**	47	30	61	43	29	33	1.30	**0.97**
Biomedical Research	12	11	**7**	3	8	2	10	16	16	0.62	**0.44**
Biology	29	1	**49**	39	41	20	29	13	11	2.64	**4.45**
Chemistry	8	1	**2**	3	11	2	6	13	16	0.37	**0.12**
Physics	3	1	**1**	2	4	3	2	17	10	0.20	**0.10**
Earth & Space	5	2	**3**	3	4	8	4	5	5	0.80	**0.60**
Eng. & Technology	5	1	**3**	3	1	2	3	4	7	0.43	**0.43**
Mathematics	2	1	**3**	1	0	1	2	2	3	0.67	**1.00**

Notes: * The activity index (AI) is obtained by dividing the share of the corresponding country or region by the share of the world.
LA: Latin America, Nig: Nigeria, Ken: Kenya, Zim: Zimbabwe, Tan: Tanzania, IvC: Ivory Coast, Zam: Zambia.
Source: Zymelman, 1990, tables I-3 and I-6.

lopsided when compared with that of the world, or Latin American countries.

There is a strong concentration of publications in clinical medicine and biology, to the particular detriment of chemistry, physics, and engineering and technology. In the case of Zimbabwe, this tendency is particularly strong with respect to biology where the country concentrates about 50% of the total number of its publications. On the other hand, although the country's absolute number of papers is very small, it practically matches the world average share of papers in clinical medicine and mathematics.

Given the small total number of publications it might be risky to extract conclusions and venture explanations about the field distribution of the papers. For historical reasons having to do with colonial dependence and the early build-up of hospitals and medical schools, research in medicine and the biological sciences tend to arise first in developing countries all over the world. Published scientific output in Zimbabwe seems to have followed the same general pattern.

3.4 Support institutions

Compared to such countries as Korea or Brazil, Zimbabwe has a very limited network of institutions to support new enterprise development and technical advance in industry (Nelson, 1993, chapters 11 and 13). Some institutions, such as the Scientific and Industrial Research and Development Centre (SIRDC) and the Zimbabwe Investment Centre (ZIC), although assigned prominent roles are not yet well established. Information about a few major institutions visited is reported below.

Harare Polytechnic

Incoming students are high school graduates that have taken English, Mathematics and three more subjects. Their programmes combine academic training with apprenticeships in industry. After four years of study and practice students are tested to become journeymen. They face excess demand for their training receiving over 100,000 applications for 7,000 positions.

So far all their graduates have found jobs. The main employers of Polytechnic graduates are the para-statal and large private sector plants. The smaller private sector plants pay lower wages and seem to prefer to train their own personnel.

The institution's budget is controlled by the Ministry of Education, and its resources come from the training levy. However, the levy is not earmarked and the Polytechnic has to bid for funds every year. They have the capacity to generate a substantial amount of funds doing applied research and selling technical services to industry, but have little incentive to do so because all revenue will go to the central government.

The Polytechnic offers a great variety of technical courses, and in some fields, such as mass communication (journalism), they have trained cadres for South Africa, Namibia and Botswana.

Most of their equipment is very old and has maintenance problems. Some has broken down and there are no resources to repair it. The Polytechnic often depends on former expatriate teachers who send them spares and other equipment to keep the school running.

Standards Association of Zimbabwe (SAZ)

The institution was formed in 1957 and incorporated in 1960; it is a not-for-profit private association governed by a General Council with representatives of government, local authorities, professional institutions and industry and commerce. In 1987 the Zimbabwe House of Assembly passed the Development Fund Levy Act, by virtue of which a levy (0.15%) is collected by the Ministry of Industry from all non-public payrolls. The Association receives a subsidy from the Levy Fund and also generates revenue from the Mark Certification programme, fees for laboratory testing, and sales of its publications.

The five principal functions of the SAZ are: (i) to encourage the use of standards of quality and to publish Zimbabwe standards, (ii) to make available laboratory facilities for the testing of goods and raw materials, (iii) to operate quality certification programmes, (iv) to operate a registration scheme for companies complying with international quality standards, and (v) to provide a library and information services.

Standards specify materials and products, codes of safe and sound practice, as well as nomenclature, methods of testing, etc. Besides improving the quality of local goods, standards are also vital in promoting the export of Zimbabwe's manufactured goods, and for the protection of consumers against inferior quality imports. Standards are prepared by technical committees including representatives from all interested parties. A standard is prepared only after an enquiry has shown that the project is desirable and worth the effort involved. The work is based on voluntary agreement and on the recognition of a community of interest among producers and consumers.

Under the certification marking scheme, goods and materials produced under proper control may bear the Association's Certification Mark. The registration scheme for companies complying with International Standards 9000–9004 (SAZS 300) is still in its beginnings with very few companies having achieved the ISO 9000 certification in Zimbabwe.

The Association has relatively new, modern facilities, and its library contains copies of all International Standard Organization (ISO) standards, British Standards (BS), South African Standards (SABS), as well as many others for public use.

Although the SAZ has a relatively long trajectory in the country, it has had limited impact when one compares its success with that of Kenya's Bureau of Standards (KBS). Probably as a result of the respect earned by its large cadre of very well qualified technical professional staff, and although a public institution, the KBS has much more influence on industry and commerce, has issued a substantial number of standards and encouraged a relatively large number of firms to adopt quality systems leading to ISO 9000 certification.

Small Enterprises Development Corporation (SEDCO)

SEDCO's mission is to serve as a centre of excellence and innovation for the development of viable small enterprises, by providing a package of financial assistance, training, counselling and related support services.

Created by a government act of 1983, and responsive to the Ministry of Industry and Commerce, it has had, so far, limited impact on the development of black entrepreneurship, in general, and in the industrial sector in particular. One of the reasons is arguably its limited capitalization. A line of credit was granted by the World Bank and opportunely used up, and the Canadian International Development Agency also collaborated providing counterpart funds to the government of Zimbabwe, as well technical assistance for management training. An expansion of the capital base to US$ 250 million was being sought.

Besides the main centres of Harare and Bulawayo it also has branch offices in Mutare and other cities in the interior. Its senior management is all Black. Its operations cover the fields of commerce, industry, construction and service.

As part of their 'development' function they offer training courses on business management in the branch offices attended by client and non-loan clients. Topics include: general business management, record keeping, production and operations management, cash management, stock control management, marketing and costing.

They also provide technical services, including, consulting assistance to small enterprises (CASE), the provision of accounting and secretarial services, the upgrading of specific skills, and the preparation of project profiles for various manufacturing industries.

In addition, they run a special Entrepreneurship Development Programme (EDP), which holds workshops of different durations, and a residential course for the examination of business plans. A majority so far have been for manufacturing projects, followed by service projects.

During the three lending periods 1989/90, 1990/91 and 1991/92, the allocation of their loans, as shown in Table 3.5, in number, varied from 47 to 57% to the commercial sector, 20 to 30% for industry, 15 to 20% for service, and 2 to 3% for construction. In terms of value lent, the allocation was: commercial 18 to 34%, industrial, 24 to 28%, service, 31 to 50% and construction, 4 to 8%.

As to the main uses to which the funds lent were applied, for the same three lending periods referred to above: working capital, 20 to 30%, machinery and equipment, 30 to 39%, motor vehicles, 13 to 44%, and land and buildings, 4 to 13%.

Table 3.5 Zimbabwe: Small Enterprises Development Corporation: sector allocation of loans, 1989/90 to 1991/92

Period	*Commerce*		*Industry*		*Service*		*Construction*	
	Numb. %	*Vol. %*	*Numb. %*	*Vol. %*	*Numb. %*	*Vol. %*	*Numb. %*	*Vol. %*
1989/90	53	25	30	28	15	42	2	5
1990/91	47	18	30	24	20	50	3	8
1991/92	57	34	20	28	18	31	3	4

Source: *SEDCO, Annual Report 1992.*

Although its lending had increased from 148 operations in 1984/85 to 543 in 1991/92, in terms of value this represented only 45 million in 1991/92. Besides its low capitalization, SEDCO faces also limitations due to the scarcity of foreign exchange necessary to import industrial machinery and other equipment, as well as the need to request collateral assets for lending to new enterprises.

Part II

Conceptual Framework and Methodology

4
Conceptualization of Technology

Acquisition of technology and the growth in technological capabilities, at the plant, firm, and country levels, may be conceived as an evolutionary process including three discernible main stages: **acquisition, operation cum attending modification (adaptation), and creation, of technology** (see Teitel, 1993, chapter 11). The way industrial firms organize internally to manage technology acquisition, operation, and change, is conditioned by several factors among which size and foreign connections seem to be of paramount importance.

4.1 Technological development as an evolutionary process

Technology acquisition

In the conception of technology adopted in this work, manufacturing technology is the **technical information** necessary to produce manufactured goods. Such information could be acquired in various ways, and it comes in embodied or disembodied form. Technology could be embodied in machinery or other plant equipment, while technical 'know-how' could be embodied in personnel with particular technical, scientific, or operational training.

Technology may be disseminated in codified form. Blueprints, product, raw materials and process specifications, and operating manuals, constitute examples of codified technical information. Or it may be part of the accumulated experience of plant workers and supervisors, and transmitted by word of mouth or example.

Whether the technical information comes in embodied or disembodied form, its acquisition is not an automatic process and it does not take place in a vacuum. In order to successfully incorporate new technical information, certain prerequisites must be met. In particular, adequate **receivers** must be in place. Qualified personnel and an appropriate organization must be available before the **transfer** of such information could occur. Prior investments in human capital and in organizational capabilities are necessary to **decode**, and, in general, to incorporate and to process the new information about manufacturing techniques, designs, or products.

The transfer of technology is consequently not a spontaneous process in which the recipient can play a passive role. Nor is technology directly available from a shelf to be acquired as an undifferentiated commodity. Instead, it has the characteristics of a special factor of production. The same technical information will be made available in different ways to different economic environments, depending on the previous existence of technology-handling capabilities – in particular, the availability of human resources and industrial infrastructure (Teitel, 1993, chapter 7).

Given the above characterization, it also follows that the transfer of new technical information will generally involve a two-way process, with interactions among **transferrers** and **receivers** of different intensity and complexity, according to the nature of the technology involved, the sophistication of the recipient's industrial facilities, and his level of technical knowledge.

Modification of technology

Experience shows that there seldom is technology acquisition, i.e. the successful absorption of manufacturing technical information, without some modification, or adaptation, of that information to local conditions. Although due to different reasons in each case, this adaptation will happen with technology transfers to both industrialized and developing countries.

In less industrialized countries, due to market size limitations, the technology must frequently be adapted to smaller size of plant, or production batch. Such adaptations are known as 'down-scaling', being quite common in the chemical and other process-centred industries.

Another frequent reason to require changes in the technology adopted, are local raw materials supplies with different quality specifications that those used in the original process. Adaptations may also be needed on account of such factors as: different tastes, changed relative prices, climatic conditions, lack of particular skills or facilities (such as repair and maintenance for example), etc. (see Teitel, 1993, chapter 8).

Whatever the reasons that make adaptation necessary, successful adaptation of a new technology to an environment different from the one for which it was originally designed and tried out, does require engineering and technical operational skills. The nature of those skills will of course vary with the nature of the technology in question, i.e. whether it is based on mechanical, electronic, chemical, or other scientific and technical principles.

While the modifications needed in the product, process or equipment, are carried out, simultaneous *creation* of new technical knowledge may take place. An adaptation may thus lead to a technological improvement which would not have occurred had the need to adapt not been present.

The creation of technology

New technologies may consist of the introduction of new products or processes as well as of the incremental improvement of existing products and their production methods. While the incremental improvements in technology may seem less spectacular, their importance should not be underestimated.

Technical changes such as the redesign of a product to improve its performance, or to add new desirable characteristics, fall in the category of what is commonly called changes in **product** technology, to differentiate from changes in the **process**, or methods of production. The latter type of changes may, or may not, include changes in the machinery and equipment used in production.

While the above distinctions (about major or incremental technical change, or between product and process technologies) may have didactic usefulness, they clearly involve arbitrary boundaries. It is quite common that when product modifications are introduced, production methods also result altered. For example, higher performance specifications may require enhanced precision involving stricter fit and tolerances which in turn demand different machining equipment or materials. Similarly, the cumulative effect of a number of small, incremental technical changes may in the end be as strong, or stronger, than that of a so-called 'major innovation'.

The nature of these technical changes is related to the technologies in use in the major branches of manufacturing. Technical activity is normally greater in some of them than in others. For example, in industrialized countries, the industries using electronic and chemical technologies have in general a much higher propensity to carry out R&D than those utilizing mechanical, or 'product specific' technologies – such as those applied in the clothing, textiles, footwear, or furniture industries (see Teitel, 1993, chapter 6).

4.2 Organizing for technology acquisition and technological change

Absorption and change of technology are activities that may occur in a deliberate and organized fashion, or they may have been left to the vagaries of entrepreneurial idiosyncracy. Whether formally or informally, such activities take place in all industrial firms even though it may sometimes be difficult to identify the agents involved in their performance or to separate operational from technological tasks.

Internal division of labour and size of firm

The extent to which a firm will be able to afford an organizational structure with differentiated and well-defined functions such as production, R&D, finance, sales, etc., will essentially be determined by its size. Size of firm is a difficult concept to measure and it has relative attributes that will vary with the nature of the environment. For example, many firms considered small in an industrialized country will be most likely considered medium or large size in a developing country. There is also considerable discussion about which is the most appropriate indicator of size; whether it should be sales, employment or assets. From the viewpoint of the present study, what matters really is the extent to which such measures are associated with a technological performance differentiating criterion, i.e. whether a small firm enjoys, or not, technical capabilities similar to those generally found in large firms.

In general, micro-enterprises and small firms cannot afford an internal organizational structure with a well defined division of labour for the major functions performed in the firm. For similar reasons, they generally lack highly trained technical personnel – specially engineers and scientists. Small firms may even be unable to hire consulting and other technical services because of their lack of awareness about the potential benefits of such services, as well as their inability to contract with them and to supervise their work (see Teitel, 1981b). Some small firms may overcome these limitations due to the presence of an owner, or manager, with special technical abilities who becomes the firm's technological 'factotum'.

On the other hand, large firms, whether private or public, and local or multinational, can afford departmental structures that are functionally entrusted with the carrying out of technology-improving activities. Subsidiaries of MNCs generally utilize the R&D and other technical services provided by facilities located at their headquarters. The centralization of such services results in further economies of scale for the MNC while it may effectively contribute to inhibit the development of a local R&D capacity. On the other hand, subsidiaries of MNCs tend to show a greater propensity to hire local scientific and engineering personnel, who may in this way acquire the requisite experience to start their own enterprises or to become more easily employable elsewhere in the same or in a related industry. Some MNCs have established certain R&D capabilities in semi-industrialized countries where sufficient scientific and engineering local talent exists. Such developments are of course a function of the level of industrial and technological development locally attained, as well as of company policy.

Research and experimental development

Industrial R&D is concerned with the purposeful pursuit and generation of new knowledge, and with its application to industrial production. R&D activities are generally carried out by units expressly entrusted with the performance of such functions, and, as noted above, only fairly large firms have R&D departments expressly set up for that purpose and possessing the critical number of personnel (scientists, engineers and technicians) fully devoted to such endeavors.[23]

As also mentioned above, *ceteris paribus*, the extent of the R&D effort will depend on the research intensity of the particular industry, and industries vary substantially on this respect. A significant correlation generally also exists between the research intensity of an industry and its employment of engineering and scientific manpower (Teitel, 1993, chapter 6).

Trouble-shooting, repair and maintenance, and quality control activities, although they may employ engineering and other technical personnel, generally, as well as in this study, are not considered part of R&D. On the other hand, product, process, or equipment improvements need not arise only as a result of the work of such expressly established R&D units. Technical change may originate in the plant floor as a result of operational, trouble-shooting, or maintenance activities, or be the result of workers' or foremen's suggestions. Many times, for lack of a facility expressly built for that purpose, the factory floor is used as pilot plant (or laboratory bench) for the testing of new designs, procedures, materials, or equipment.

Industrial engineering

To a greater or lesser extent, most technical change generally entails modifications in hardware of one type or another. There is, however, a type of technical change which relies mostly on improvements in the organization of production and is the result of the work of industrial engineers. Industrial engineering skills could be used in all industries; they are thus 'universal' in nature. The function of industrial engineers is the continuous improvement of productivity. They follow the motto that everything can be improved.

Industrial engineering may be defined as: 'the systematic search of better ways of doing things, leading to productivity increases by means of changes in the time and spatial sequence of manufacturing and auxiliary operations, and in the methods and organization of production'. That is, the emphasis is not on new or better hardware, but on eliminating unnecessary operations, reducing delays between operations, and bringing down inventories.

To achieve their objectives, industrial engineers apply a variety of mathematical, statistical, and managerial techniques, including: operations research (encompassing various optimization algorithms), time and motion studies, layout and materials-handling analysis, work-study methods, work sampling, job evaluation, time-piece and incentive remuneration methods, and organizational analysis.

Other engineering functions

As already indicated above, *ceteris paribus*, the type of engineering manpower to be found in a manufacturing firm, and the engineering functions performed, will primarily depend on the technical characteristics of the industry it belongs to. Metalworking, electro-mechanical, and electronic consumer products industries will generally have engineering departments engaged in product design, and in the improvement of manufacturing methods. Chemical and food-processing industries will emphasize process-optimization engineering departments.

The technologies in use will determine the skill composition of the engineering work-force – i.e. mechanical, electrical, chemical, etc.

There will also be cases requiring engineering skills only applicable to specific industries such as: textile, petroleum, and automotive engineers. On the other hand, as noted above, industrial engineering is the only type of engineering skill of general, or universal, applicability across industries (Teitel, 1993, chapter 6).

5
Methodology

Below are briefly described the field work carried out, the sample of firms interviewed, and the main hypotheses applied in the preparation of the questionnaire used to survey manufacturing firms from selected industry sectors in Zimbabwe. The main purpose of this survey was to assess the technological and skill capabilities in these sectors as well as their potential needs for support services and other technical assistance.

5.1 Field work

The field work on which this study is based involved interviews in a selected sample of manufacturing establishments, using a special, previously designed and tested, questionnaire (Appendix A), as well as the visit of selected industrial support institutions.

The interviews were carried out by two teams of researchers, each one including a senior international consultant plus a local investigator affiliated with the University of Zimbabwe. The composition of the teams was changed every couple of days by rotating the international consultants to afford greater exposure and interaction among the team members as well as to homogenize interviewing procedures.

Plant visits were an integral part of the interview process, and the plant was the focus of the data collection exercise. In some cases this was not easy to enforce because some of the firms studied belonged to holding groups and/or operated more than one plant, on occasion, in the same premises.

The interviews required from one and a half to three and a half hours approximately, with less time generally needed for the smaller firms –

although loquacity of the respondents was also a significant variable affecting the duration of the interview. The interviews were carried out with the owner, or general manager (managing director), in the case of small firms, and with the general manager (managing director) and plant manager in larger firms. These people some times delegated meeting with the interviewing team to others (for example, the financial manager or administrative manager, instead of the general manager, or the production manager instead of the plant manager), but the interviews were in most cases carried out at the two desired levels: general management and plant management.

Three-quarters of the interviews were programmed for enterprises located in the capital city of Harare and its suburbs, and one-quarter for firms in the city of Bulawayo. A total of 41 firm interviews were successfully carried out, 31 in Harare and 10 in Bulawayo.

In addition to the firms, the following institutions were visited: Standards Association of Zimbabwe (SAZ), Harare Polytechnic, Faculty of Engineering of the University of Zimbabwe, Zimbabwe Institute of Engineers, Small Enterprises Development Corporation (SEDCO), Scientific and Industrial Research and Development Centre (SIRDC), and Zimbabwe Investment Centre (ZIC).

5.2 Sample

The 41 firms for our survey were randomly extracted from a larger (panel) sample of 200 preselected firms and stratified to reflect industry composition, location, size, ownership structure, and ethnic background.[24] Because small firms had more problems arranging the interviews and sticking to the agreed upon schedule (even after reconfirming the appointments), large firms (with employment greater than 100 employees), were somewhat over-represented, while smaller, black-owned firms, were under-represented among the enterprises actually interviewed.

Industry breakdown

The sample included 13 firms (31.7%) from the **food and beverages** industry, 10 (24.4%) each from the **textile and garments** and **wood and furniture** industries, and 8 (19.5%) from the **metalworking** industry. Table 5.1 shows the industry composition of the sample.

Due to the need to meet the various criteria stated above, the sectoral distribution of firms in the sample does not necessarily correspond to

Table 5.1 Zimbabwe: industry composition of survey sample

Food		*Textile*		*Wood*		*Metalworking*		*Total*	
Num.	%	*Num.*	%	*Num.*	%	*Num.*	%	*Num.*	%
13	31.7	10	24.4	10	24.4	8	19.5	41	100

the industrial composition of the manufacturing sector. According to the 1989 Census, food, beverages and tobacco accounted for 30% of manufacturing output; textiles, clothing and footwear, for 18%, wood and furniture for 3%, and metals, metal products and transport equipment for 24% (World Bank, 1993b, p. 18).

Location

Thirty-one (75.6%) of the firms interviewed for the study were located in Harare and its surroundings, and the rest (10 or 24.4%) in Bulawayo. As noted above, these two cities, and their suburbs, together account for approximately three-quarters of the country's manufacturing output. The locational distribution of the firms in the sample is shown in Table 5.2, and it can be seen that, as desired, it reflects quite well the underlying locational distribution of industry in the country.

Size

The size classification adopted takes into account Zimbabwe's relatively more developed manufacturing sector. This includes, *inter alia*, a larger average size of firm and a greater number of 'large' firms than would have been the case in other Sub-Saharan Africa countries.[25] Thus firms were classified as follows: **small**, those employing up to 10 workers, **medium**, those employing between 11 and 100, and **large** those employing more than 100 workers.

Table 5.2 Zimbabwe: location distribution of firms in the sample

Harare		*Bulawayo*		*Total*	
Number	%	*Number*	%	*Number*	%
31	75.6	10	24.4	41	100

Table 5.3 Zimbabwe: size composition of sample

Small		*Medium*		*Large*	
Number	%	*Number*	%	*Number*	%
4	9.8	14	34.1	23	56.1

In Table 5.3, the size composition of the sample is shown. Compared to overall manufacturing, the 'large' subsector is over-represented, and the number of 'small' firms is quite limited; reasons accounting for their under-representation were mentioned above.

Ownership and ethnic origin

As shown in Table 5.4, in the sample, twenty-nine firms (70.7%) are private and locally-owned, two (4.9%) are private of foreign ownership, three (7.3%) are private joint-ventures with foreigners, five (12.2%) are joint ventures of the state with foreigners, one is a state and local private owners venture, and one is a state and foreign private owners venture. Compared to the manufacturing sector as a whole, there is a slight over-representation of the state sector in the sample.

In Table 5.5 the ethnic origin of the firms in the sample is shown cross-classified by industry.

Table 5.4 Zimbabwe: ownership structure of the sample

Priv(Z)		*Priv(F)*		*Priv(Z&F)*		*St&P(Z)*		*St&P(F)*		*State*		*Total*	
No.	*%*	*No.*	*%*	*No.*	*%*	*No.*	*%*	*No.*	*%*	*No.*	*%*	*No.*	*%*
29	70.7	2	4.9	3	7.3	1	2.4	5	12.2	1	2.4	41	100

Table 5.5 Zimbabwe: ethnic background of the sample

Ind.	*Asian*		*European*		*African*		*Other**		*N.A.*	*Total*
	No.	*%*	*No.*	*%*	*No.*	*%*	*No.*	*%*	*No.*	*No.*
Food	1	8.3	7	58.3	1	8.3	3	25.0	1	13
Tex.	1	11.1	5	55.5	3	33.3	0	0	1	10
Wood	0	0	6	75.0	2	25.0	0	0	2	10
Met.	0	0	4	57.1	0	0	3	42.8	1	8
Total	**2**	**5.5**	**22**	**61.1**	**6**	**16.7**	**6**	**16.7**	**5**	**41**

Note: *Includes firms with mixed background, public corporations and state firms.

As can be seen from Table 5.5, more than half the firms for which data was available in the four industries are owned by Europeans, or people with European ethnic background, while only six had African (black) ethnic background.

5.3 Survey hypotheses

The survey methodology applied in this study has advantages and limitations. On the positive side, and due to the detailed nature of the information elicited in the interviews and plant visits, a qualitatively rich account of the technological capabilities and shortcomings, as well as detailed data on technical skills, in the manufacturing plants visited could be obtained. The main limitation of the method resides in the relatively small size of the sample that it is possible to utilize, precisely because of the detail and extension of the interviews conducted. The trade-off is thus between statistical validation and the qualitative richness of the information gathered. To compensate in part for this situation, use is also made of selected technology-related data available from a somewhat larger (panel) sample.

The interviews were carried out following a pre-designed questionnaire copy of which is included as Appendix A. The questionnaire was designed to test an evolutionary conception of technological development as indicated in the previous chapter,[26] as well as specific hypotheses at the country, sector, and firm levels, some of which are spelled out below.

Besides firm and entrepreneur background information, and data on skilled personnel, the questionnaire comprises five main blocks of questions, on: (i) acquisition of technology, (ii) technical operating capabilities, (iii) technological functions and their organization, (iv) plans for technological development, and (v) technical skills.

The following are some of the main hypotheses, corresponding to the country, industry, and firm levels that underlie the questionnaire's design and the formulation of specific questions.

Country characteristics

Here we include hypotheses resulting from the country's historical background, industrialization process followed, and recent economic policy changes as they affect industrial and technological development.

(a) Together with other Sub-Saharan Africa countries, Zimbabwe is in the initial stages of its technological development process, characterized by a predominance of the **acquisition** phase, with limited attainments in **modification** (adaptation), and/or **creation**, of technology.
(b) However, for historical and other reasons noted in Chapters 1 and 2, Zimbabwe's manufacturing capabilities are substantially greater than those of other Sub-Saharan Africa countries.
(c) Given the level of industrialization it has attained, the structure of its manufacturing sector, although characterized by a concentration on low-skill activities, also shows a greater degree of diversification than other Sub-Saharan Africa countries, including a substantial metal products sector.
(d) Because of the country's industrialization history, market orientation is substantially domestic, although the proportion of manufacturing exports is greater than in other Sub-Saharan Africa countries.
(e) Given its low level of technological development, most technology used in Zimbabwe's manufacturing plants will be of foreign origin, and although some local production of machinery exists, the source of most of its industrial equipment is also expected to be foreign.
(f) Due to the effects of structural adjustment and trade liberalization policies recently adopted, it is to be expected that some firms, especially in those industries hardest hit by the new policies, will not be operating at full capacity.
(g) As part of their efforts to adapt to the new economic environment by improving their competitiveness, it is to be expected that some firms would be introducing technical changes in production to increase productivity and improve quality.

Industry characteristics

Manufacturing industries tend to differ in skill composition and the technologies they use. These differences are also to be reflected in the four industries selected for our study.

(a) Due to public health requirements, plants in the **food and beverages** industry will have stricter quality control requirements and utilize laboratories (their own and/or external ones), particularly for microbiological and other chemical controls.

(b) Because of precision and other technical quality requirements characteristic of many of its products, the **metalworking** industry will hire a relatively higher proportion of technical personnel – both engineers and technicians.
(c) Due to product quality and process requirements, as a rule, **food and beverages**, and **metalworking**, will be more skill intensive than **textiles and clothing**, and **wood and furniture**.
(d) Process-centred plants, to be found mostly in the **food and beverages**, and **textile** sectors, will tend to pay special attention to maintaining their equipment in continuous operation by minimizing breakdowns or quality control problems.
(e) Plants operating to order, to be found mostly in **metalworking**, **furniture**, and **clothing**, will be more concerned with meeting customers' product specifications.

Firm characteristics

The main distinctions among firms with respect to technological behaviour stem from size (between large and small), and ownership (between foreign and local) differences.

(a) Because of internal division of labour limitations inherent to their reduced size (see Chapter 4.2), small size enterprises, which constitute a substantial part of Zimbabwe's manufacturing industry, tend to have a low level of technological capability.
(b) *Ceteris paribus*, local firms will be at a significant disadvantage *vis-à-vis* subsidiaries of MNCs in terms of access to technology and technical assistance.
(c) As a rule, subsidiaries of MNCs do not engage in major R&D activities in countries such as Zimbabwe because their corporate policies preclude such efforts in places without an adequate complement of science and technology infrastructure and support services, and a substantial supply of highly qualified R&D personnel.
(d) Because of size-imposed restrictions in their internal division of labour, small firms will concentrate a number of managerial functions in one or two executives, and will tend to have limited access to high-level technical personnel.
(e) Larger firms can afford a functional organization including individual heads for the main departments: sales (or marketing), finance, production, engineering, personnel (or industrial relations), etc., as well as access to high and intermediate-level technical personnel.

(f) Limited codification of production and product technical information is expected to be found. The extent of codification will be related to such factors as size and ownership. Larger firms, and those associated with MNCs will, as a rule, have more comprehensive systems of codified information and rely less on the transmission of know-how via word of mouth, and/or on the experience accumulated by senior operating personnel.

Part III

Manufacturing Technology Acquisition and Operation

6
Acquisition of Technology

As conceptualized in this study (see Chapter 4), **acquisition** is the initial stage in the process of technological development. Moreover, we have also learned in the last chapter about the hypotheses underlying construction of the questionnaire used for the firm interviews. It was assumed that most of the technology was of foreign origin and transferred by various mechanisms to Zimbabwe. We also noted systematic characteristics that may affect technological performance at the industry and firm level.

The questions asked about technology acquisition fall within three blocks: (a) about the origin of the firm, (b) about the sources and characteristics of its equipment, and (c) about the use of foreign disembodied technology.

6.1 Origin of the firm and initial studies

Many of the Zimbabwe firms surveyed are quite old and they also make use of old machinery. In a number of them, established more than 25 years ago, institutional memory is limited and current management does not know much about the firm's origin, as expressed by vague statements such as 'it began as a small operation'. Other firms, mainly the newer ones, recall their origin. Twenty-two, or slightly more than half the firms interviewed, reported that they began as small operations under the current ownership; 12 (less than one-third) were acquired as on-going concerns by the current owners (most of whom believed that these also began as very small operations); only four started as full-scale plants (one of these was a joint venture); and three plants did not provide any information about their origin.

The entrepreneurs that began the firms frequently had production experience in their fields. The use of second-hand machines to start the firm has been quite common, even in larger plants. For example, in one case, six firms each contributed used equipment to form a new company. In this case, most of the equipment had been previously operated in South Africa, and later additions have followed the same pattern. In another case, used equipment was brought from England to start the operation.

In most cases no formal feasibility studies were carried out before starting the business. Twenty-eight plants surveyed reported no such studies, and only five reported studies. The rest (eight) had no available information. Among the few firms that had done feasibility studies, in two they were done by the entrepreneur himself, in two by in-house personnel, and in two by the foreign partner, headquarters, or with their participation. No firm reported having had any help from outside the firm or its partners in this respect.

In some cases, the entrepreneurs 'followed their intuition', and in others they simply felt that they knew the business well enough to be able to succeed without a formal evaluation of their investment. Since many of the machines were second-hand and represented a low investment expense, the risk involved was not deemed to be too high. In other cases, when the investment required was large, the market, or access to the principal input, was somehow secured.

Several firms with substantial fixed capital investments were established to use large domestic input supplies that would have been wasted otherwise, such as milk and pork by-products. In these cases the manufacturing plants were created to fulfil well-known demands of specific constituencies, and there was no critical need to do a formal project appraisal. In those cases where firms have changed hands since they were established, buyers have generally been more careful and current managers informed that project evaluation studies were made at the time of the purchase.

The frequency of second-hand start-up machinery reflects the financial limitations of many budding entrepreneurs, and the tight foreign exchange constraint experienced by the country since 1965 and until recent times. During the early 1980s, funding and foreign exchange were made available to import machines from the former communist block at significantly lower prices than those of similar equipment made in the West or the Far East. In one case, a plant was started with new equipment from an Eastern European country, but when it expanded it used second-hand equipment from those other regions because of difficulties experienced with the original equipment.

6.2 Source and characteristics of the equipment

As noted above, a key characteristic of the machinery and equipment currently used in Zimbabwe is its old age. Another, generally correlated with age, is the relatively low level of technology embodied in those machines.

Equipment selection

The *choice of machinery to begin operations* was most frequently made by the entrepreneur him/herself (18 cases). In-house personnel, local consultants, foreign consultants and foreign partner/headquarters each chose the machines in two cases, while 'other' was mentioned seven times. The most frequent 'other' category was 'machinery available in the local market'. In the cases of *subsequent expansions*, participation of the entrepreneur in this role declined to nine cases, while 'other' rose to 16, in-house personnel to six, and local consultants to four. Expatriate consultants remained at two and foreign partner/headquarters dropped to one. These results seem to indicate that once the plants were operating, the choice of machinery for expansion was more influenced by technical personnel within the plants, and by what was available in the local market, than when the plants were originally established.

The role played by the firm in specifying its initial equipment appears unclear. About one-half of the firms (19) could not answer this question. Of the rest, 10 claimed that they had not played any role in this choice, and in two cases the equipment had been specified by headquarters. Only about a quarter of the firms (10) claimed to have played a strong role in specifying their original equipment: in six cases the firm fully determined the type of equipment it wanted, and in four others they interfaced with suppliers to decide what to get.

The age of many plants and the lack of institutional memory is reflected in the fact that only 17 plants could provide information about the *method* used to locate the original equipment. About half of them had decided what machines, and how to buy them, based on the entrepreneur's familiarity with the manufacturing process. They had experience in production and a pretty good idea of what they wanted. In other cases, the machinery and equipment sources were: foreign

Table 6.1 Zimbabwe: methods used in locating plant equipment (number of times each method was used)

Sector	*Trip abroad*	*Catalogue*	*Distributor*	*Foreign partners*	*Experience*	*Other**
Food	0	0	1	1	2	1
Textile	0	1	0	0	3	2
Wood	1	1	0	0	2	0
Metal	0	0	0	1	1	0
Total	**1**	**2**	**1**	**2**	**8**	**3**

Notes: *Includes: one instance of machine provided by local Muslim community, one of machines discarded by large factory, and one of machines personally seen at work.

partners (two), local distributors (one), catalogues (two), trip abroad (one), discarded old machines from a larger plant (one), and, in one special case, the Muslim community provided a fellow congregant, at no cost, the second-hand equipment he needed to be in business (see Table 6.1).

Age of the machinery and equipment

Unable to obtain information about all pieces of equipment in all the plants visited, we inquired about the age and characteristics of the **three main** machines or pieces of equipment in each plant visited. In this way information about 120 pieces of equipment was obtained. In many cases the respondents could provide only approximate answers to questions about the age of the equipment in use. Because of this, the answers tend to concentrate around round numbers. The results show that only 20% of the equipment is less than 10 years old. However, 30.8% was reported to be 10 years old. At the other end, over 28% of the main pieces of equipment are 20 or more years old, over 17% exceed 30, and over 13% exceed 35 years of age. The mean age reported was 16.2 years, and the standard deviation was 11.8 years. These results indicate that a large part of the main pieces of equipment used in Zimbabwe's manufacturing plants surveyed are quite old.

Equipment sources

The machinery in use in the sampled firms comes from a number of countries. About 30% of the principal items of equipment originated in the United States and 19% in Germany (see Table 6.2). Interestingly, about 12% comes from Zimbabwe itself, reflecting the relatively high level of metal fabrication development attained by the country due to

Table 6.2 Zimbabwe: sources of principal items of equipment

Source	*Frequency*	*Percent*	*Cumulative frequency*	*Cumulative percent*
France	1	0.8	1	0.8
Germany	24	18.8	25	19.5
Holland	1	0.8	26	20.3
Italy	5	3.9	31	24.2
Japan	10	7.8	41	32.0
Norway	1	0.8	42	32.8
South Africa	8	6.2	50	39.1
Sweden	1	0.8	51	41.4
Switzerland	2	1.6	53	52.3
UK	14	10.9	67	82.0
USA	38	29.7	105	93.7
Zimbabwe	15	11.7	120	100.0
not available	8	6.2	128	

Source: Sample survey.

its history of economic isolation and foreign exchange constraints. The rest of the main items come from the United Kingdom (11%), Japan (8%), South Africa (6%), and other Western European countries. Firms also reported having bought smaller pieces of equipment from such countries as Bulgaria, India and Taiwan.

New or used equipment?

The survey showed that 68.7% of the principal machines and equipment were bought new, 30% used, and no information was available about 1.3%. As noted above, many industrial plants have relied on used equipment in Zimbabwe, and the second-hand figure is over three times higher than the one found in a similar survey done in Kenya (8.9%) (see Teitel, 1994b).

As noted above, second-hand machines have been attractive because of financial and foreign exchange constraints, and it appears that Zimbabwean entrepreneurs have developed their own strategies and skills for purchasing and importing used equipment. For example, a metalworking entrepreneur explained that he would be very reluctant to buy a second-hand machine from a successful firm, preferring instead to buy from bankrupt firms. He reasoned that buying from a bankrupt firm he was less likely to buy a machine whose owners wanted to discard it because of excessive wear or maintenance problems.

Satisfaction with current equipment

To find out whether managers were satisfied with their current plant, the survey asked: if they were to buy new equipment to produce the same products, would they buy the same or different machinery? The answers to this question were quite revealing. Only five of the 41 firms were satisfied with their technology and production capacity, and would buy again the same equipment they have now. The answers of the other firms were quite variegated and reflect a need for change and a desire for technological improvement. Six plants would have liked to have the same productive capacity but newer technology. Sixteen firms would prefer to have a plant that would allow them to produce larger quantities with new technologies, while seven more would also like to be able to produce larger quantities but with the same technology. In six cases, the firms would like other type of equipment. These cases include a firm that is already undertaking a very drastic change in both technology and productive capacity, and several more that would like to change their output mix, to add or to eliminate production lines.

The reasons given to justify the changes desired in production are as follows: in 27 cases management wanted to lower costs and increase productivity, in 24 cases to increase quality, and in seven cases to introduce new production lines or to modify the production mix. In 18 cases there was a desire to improve productivity, and in 11 just to increase output.

These results are a reflection of the history of manufacturing in Zimbabwe, the policy changes currently experienced, and the expected changes in the international environment. It is clear that most firms would like to have a plant with different equipment from what they now have. In some cases, the desired changes are small, for instance, a more versatile sewing machine, but in others, these changes are more drastic. The desire not just to lower costs and increase productivity, but also to improve quality is quite noticeable. This is a significant departure from the past, when the main challenge for manufacturing entrepreneurs was to produce for the protected domestic market. As Zimbabwe's economy opens to international competition, cost reductions and quality improvements are seen as indispensable to succeed.

However, the desire for new technology is not necessarily for advanced technology, and in cases it reflects the needs imposed by the low level of training of the labour force. Managers frequently indicated that they prefer better and newer technologies to the very old ones they were using, but made it clear that they did not intend to bring up-to-date machines requiring very high skills to service and repair. In other cases, the reason why they wanted newer technology was to 'de-skill' the production process. It was expected that in the desired new machines workers' skills would have a lower impact on the quality of the final product.[27]

Technical start-up problems

Since there is limited institutional memory about the start-up period of many plants, little is known about the kind of problems they may have faced during their earlier stages (see Table 6.3). Information of 30 plants is available about whether they had problems reaching design capacity. Of these, only four faced difficulties, and only one of them could not solve the problem satisfactorily. With regard to meeting product specifications, only two of 25 plants had problems, and both solved them. Similarly, three of 25 had problems meeting process specifications, and all solved them. Equipment breakdowns were a more frequent source of start-up problems as six of 26 plants faced them; moreover, two of the six plants had not been able to overcome this problem. Other start-up problems were reported by four of 26 plants, and only one could not take care of them.

Table 6.3 Zimbabwe: technical start-up problems encountered

Sector	*Reach design capacity*	*Meet product specif.*	*High reject rate*	*Meet process specif.*	*Equipment breakdown*	*Other*	*Total*
Food	2	2	3	1	1	2	11
Textile	1	0	1	0	3	1	6
Wood	0	0	0	2	1	1	4
Metal	1	0	1	0	1	0	3
Firms w/problem	4 (9.8%)	2 (4.9%)	5 (12.2%)	3 (7.3%)	6 (14.6%)	4 (9.8%)	24
Firms without problem	26	23	20	22	20	22	–
Firms without information	11	16	16	16	15	15	–

Source: Sample survey.

The data show that among the plants with institutional memory about start-up problems, the most frequent one was **equipment breakdowns**, a fact consistent with the old age of the machinery and equipment used. Interestingly, in all but four cases, start-up problems were taken care within the plant, or with minor outside technical support. Two of these four cases had to do with equipment breakdowns related to the old age of the equipment. The other two failures were in the same plant that had not been able to reach design capacity and also had a high proportion of rejects.

These results indicate that Zimbabwean entrepreneurs in our sample have generally been successful in debugging their plants following start up, a result reflecting the presence, and/or eventual acquisition, of significant mechanical skills for fixing old machines and keeping them running.

6.3 Use of disembodied foreign technology

While most machinery is imported, manufacturing in Zimbabwe makes very little use of foreign disembodied industrial technology, including very few licences and technical assistance agreements. The larger sample (panel) data indicate that 27 firms (13.5%) use foreign licences, and 25 firms (12.5%) receive technical assistance (see Table 6.4). Similarly, the number of expatriates in Zimbabwean manufacturing is quite small by Sub-Saharan Africa standards. The same table shows only 42 expatriates, mostly in the **wood and furniture** and **food and beverages** sectors. They worked in only 20 firms, i.e. 10% of the total sample. It is worthwhile noting that a large majority of them (30 out of 42, or 71%) were employed in technical jobs.

Table 6.4 Zimbabwe: use of foreign licences, technical assistance, and expatriates (total and by sector) (panel sample)

Sector	*Licences*		*Techn. assistance*		*Expatriates*	
	Number	*%*	*Number*	*%*	*Number*	*%*
Food	10	37.0	9	36.0	12	28.6
Textile	13	48.1	9	36.0	2	4.8
Wood	0	0.0	2	8.0	22	52.4
Metal.	4	14.8	5	20.0	6	14.2
Total	**27**	**100.0**	**25**	**100.0**	**42**	**100.0**

Source: Panel sample.

The results of our survey are similar to those of the larger panel sample. Only nine of the 41 firms use any foreign licences, and five have technical assistance contracts. Five of the plants using licences were in **food and beverages**, where scientific knowledge is necessary to achieve and maintain high levels of quality. The other cases were in **textiles and apparel** and **metal manufacturing**, each with two firms. With respect to technical assistance agreements, three of the firms with such contracts were in **food and beverages** and one each in **textile and clothing** and **metalworking**. The **wood and furniture** firms in our sample were not recipients of any of the three forms of disembodied foreign technology (see Table 6.5).

Of the 10 firms that used licences or received technical assistance, five paid royalties and one a fee for service. However, nine of the 10 firms also had other arrangements to receive technical aid. Several plants were taking advantage of various free technical assistance services provided by donor countries or non-governmental organizations (NGOs), or had sui-generis arrangements such as those of a company that had foreign partners who supplied technical assistance free of charge expecting to be repaid by sharing in higher profits.

A frequent form of technical assistance was that provided by suppliers of sophisticated equipment to repair and overhaul those machines. Annual overhauls of large expensive machines have been done with the help of technical personnel coming to Zimbabwe for that particular purpose.

Among the firms that received technical assistance or used licences, three reported that the agreement had tie-in clauses for imports of materials or components, and five reported the existence of export restrictions.[28] However, several of the export restrictions applied only

Table 6.5 Zimbabwe: use of foreign licences, technical assistance and expatriates in sample firms (number)

Sector	*Licences*	*Technical assistance*	*Expatriates*
Food	5	3	8
Textile	2	1	1
Wood	0	0	0
Metal.	2	1	6
Total	9	5	**15**

Source: Sample survey.

to some markets, and some plants with export restriction clauses in their licence or technical assistance contracts were nevertheless exporting. In other cases, particularly exports of apparel and furniture, foreign technical assistance was an enabling factor to export.

The number of plants that had used technical assistance or external support to instal and debug equipment was significantly larger than the number that were using production licences and technical assistance. In total, 18 of the 41 firms (19.5%) surveyed had used some form of support to instal new equipment. The largest number (nine) was found in **food and beverages**, followed by five in **metalworking**. Only one plant in **textiles and apparel** (a textile plant that acquired a new stamping machine), and three in **wood and furniture** had used such services.

The most frequent form of technical assistance used in installing new equipment has been the supply of technical personnel from the supplier that have helped with the installation (14 cases). Personnel training to operate the equipment occurred in seven cases, five of which involved training abroad.

On the whole, the above data shows limited use of foreign disembodied technology, in its main forms or channels, and once more seems to reflect the isolation the country experienced after UDI, and the strong self-sufficiency it helped to develop among local industrial entrepreneurs.

7
Plant Technical Operating Capabilities

The test of successful acquisition of external industrial technology obviously resides in the ability to utilize the technical knowledge and equipment so obtained in the daily operations of the manufacturing plant. To ascertain the level of technical capability in plant operations the questionnaire included the following blocks of questions: (a) of a general nature, including upkeep, capacity utilization, etc., (b) about trouble-shooting capabilities, (c) quality control, (d) repair and maintenance, (e) industrial engineering capabilities, and (f) on safety and environmental control.

7.1 General

Plant facilities visited were in most cases adequate. Many of the buildings looked clean and well kept while the plants seemed to be well organized. However, in about 25% of the plants visited we found deficiencies ranging from lack of basic facilities to very crowded conditions, poor layouts and disregard for safety and cleanliness. Major deficiencies were found in a few small enterprises that have not been able to accumulate significant amounts of capital and do not have access to credit. In these cases, the premises had not always been built for manufacturing, and tended to be quite crowded. The main safety problems mostly related to lack of proper ventilation and protection equipment or to failure to use them.

As noted, one of the main characteristics of the plants visited was the old age of their equipment. The survey revealed that of 36 plants for which data were available, 20 had mostly old equipment (15 years and older), and 12 had a mix of old and new. Only four had mostly new equipment (five years or less). The old equipment was frequently

Table 7.1 Zimbabwe: distribution of firms in the sample by level of capacity utilization

Capacity utilization (%)	*Frequency*	*Percent*
Up to 50%	11	26.8
51–75%	12	29.3
76–100%	18	43.9

very old, (over 20 years of age), and it was common to find machines that were over 30 and 40 years old. The oldest machine in use was 91 years old!

Capacity utilization was found to vary substantially. This is not surprising since the manufacturing sector is experiencing a significant adjustment due to changes in domestic policies and in the external economic environment. According to management, about one-quarter of the plants visited were operating at 50% capacity or less. About 30% operated at between 51 and 75% capacity, 12% between 75 and 89% capacity, and 32% at 90% capacity or higher. (See Table 7.1.)

Capacity utilization varied significantly by sectors. **Wood and furniture** plants were operating at quite high levels. Eight of the 10 plants visited were operating at 90% capacity or higher (including seven at full capacity). In **textiles and clothing**: three of nine plants ran at 90% capacity or higher, and two at 50% or less, three at between 51 and 75%, and two at between 76 and 89%. **Food and beverages** had the lowest rates of capacity utilization: six of 13 plants operated at 50% capacity or less, and only two at 90% or higher. **Metal manufacturing** plants operated at between 40 and 85% capacity.

The above-noted sectoral differences in capacity utilization rates were probably a reflection of the following recent trends: (i) the export success of the **wood and furniture** sector, (ii) the problems faced by some **food** plants due to input supplies scarcities, (iii) the low levels of demand for machines, equipment and consumer durable goods produced by the **metal manufacturing** sector caused by the country's recession, and (iv) the downsizing of some **textiles and clothing** plants whose exports have declined due increasing costs, and whose domestic demand shrunk because of the recession and import competition.

As is to be expected in plants with very old machinery, where fixed capital costs are very low, one shift operation was the norm. In these cases, the higher night wage and electricity costs more than counterbalance the savings from using the fixed capital more intensively.[29] Of 35 reporting plants, only six (17%) operated more than one shift. However, even in those cases, only some sections and machines were operating more than one shift. Multiple shifts operation tended to be the result of unbalanced production processes, or of the need to respond to seasonal peaks in demand. In several plants operating more than one shift, this was considered to be a temporary situation and measures were being taken to correct plant unbalances in order to operate only one shift in the future.

Plant unbalances and lack of demand resulted in most plants having idle machines. However, important sectoral differences also obtained in this area. All 10 plants for which information was available in the **food and beverages** sector had idle machines. In both **textiles and clothing** and the **metal manufacturing** sector, only one out of 10 plants did not have idle machines. In contrast, four of eight **wood and furniture** plants did not have idle machines. Among the four plants with idle machines, one was in that predicament because it had grown too fast and did not have enough space to put all the machines to use, and another had machines idle because of lack of materials supply, not because of lack of demand.

7.2 Troubleshooting

Equipment breakdowns was the most frequent and most important operating problem experienced in the manufacturing plants visited. Indeed, three-quarters of the plants reported frequent breakdowns. Breakdowns seem to be strongly related to the old average age of the machinery and equipment noted before. *Bottlenecks in the production process* was the second most mentioned problem, and was present in about one-third of the plants. The presence of bottlenecks is likely to be a reflection of the long history of foreign exchange constraints experienced by the country. Serious *quality control* and *pollution control* problems were reported by nine plants each. *Safety* problems were mentioned in six cases, and *other problems* in five instances. (See Table 7.2.)

Equipment breakdowns are widespread across the four manufacturing sectors surveyed. Quality control and safety problems are also evenly distributed across sectors. However, *pollution* and *bottleneck* problems are concentrated in the wood and furniture sector. Indeed, two-thirds of the plants with pollution problems and six of 13 that had bottlenecks were in that sector.

Table 7.2 Zimbabwe: trouble-shooting problems reported by sample firms

Problem	*Frequency*	*Percent*
1. Equipment breakdowns	31	75.6
2. Bottlenecks in production process	13	31.7
3. Serious quality control problems	9	22.0
4. Pollution control problems	9	22.0
5. Safety problems (accidents)	6	14.6
6. Other problems*	5	12.2

Note: *Most of these were related to power supply. One firm complained of malfunctioning of equipment and electrical control problems, two others of electricity supply. There were also complaints about finishing and about subcontracted repairs.

The results of the survey show that, with differences depending on their repair and maintenance capabilities, most plants have found ways to cope with frequent equipment breakdowns (see Table 7.4 and below). In some cases, breakdowns disturbing the production process have become such a frequent occurrence, that dealing with them has become the plant manager's main task. Undoubtedly, Zimbabwean manufacturers have developed substantial capabilities to keep their old machines going.

7.3 Quality control

Since UDI and until the economic policy changes of the 1990s, quality control was not an important consideration for those engaged in manufacturing production as the local market absorbed almost any product. All plants visited have some quality control system, although in most cases (25 out of 41 or 61%) it is simply a visual and practical inspection. Laboratory tests and the use of national or international standards were found in 16 plants.[30] This group consists mainly of plants in the food sector and others that export.

Virtually all firms carry out some form of quality control of the final product (see Table 7.3). More than 80% of the plants surveyed also carry out quality control of purchased raw materials and components, and during the production process. The decision as to when (at what point in the manufacturing process) to do quality control depends on the type of process and on market conditions. For instance, in furniture plants, visual quality control is applied to parts in process to make sure that scratches and other imperfections are corrected before the

Table 7.3 Zimbabwe: Levels of quality control in sample firms

Existence of quality control for:	*Number*	*Proportion (%)*
1. Final product	39	97.5
2. Process(es)	34	85.0
3. Raw materials and components purchases	33	82.5
4. Materials and components in process	25	62.5

furniture is assembled. In cases where there is limited supply of raw materials or intermediate products, management has to take what is on the market, and does not apply any quality control of purchased inputs.

Rejection rates could be a useful quality indicator. All plants visited were quite concerned about consumer rejection rates and all argued that they were quite low. In many cases they explained that they had developed systems to avoid faulty products reaching the consumer. The concern about consumer rejection was greater in the **food and beverages** industry and among exporting firms. Some plants in the food industry indicated that their rejects were not related to health or sanitary issues, but to colour or taste. In this sector, management was also concerned about rejections resulting from mishandling after the products had left the plant – in particular, exceeding shelf-life and storage at improper temperature.

The recent economic opening was reported to have contributed to a lower rejection rate in some plants because of increased availability of some inputs. In the **clothing** sector, many of the rejections were blamed by management on the low quality of the cloth available in the local market. The old age of the machinery and equipment could also be a cause of low-quality products and high rejection rates. For example, an old foundry was found to produce castings that had to be redone in 10% of the cases; old sewing machines' stitches were not all of the same length, thus could not meet international market standards.

7.4 Maintenance and repair

This function is crucial in Zimbabwe's manufacturing sector because, as indicated above, the average age of the equipment is quite old due to past foreign exchange restrictions and the limited supply of machinery of local manufacture. As also noted, skills to repair and maintain the plant equipment had to be developed, in particular during the UDI period, and this produced an unintended positive externality for other types of related metalworking activities.

Only three firms were totally dependent on outside help for their repair and maintenance work, while at the other end, six firms had a complete in-house service. As shown in Table 7.4, 20 firms (49%) had the capacity to not only oil and replace accessible parts, but were also able to strip most machines and to replace standard parts or get spares made to order. On the other hand, no firm practices complete **preventive maintenance** which could be considered as the highest level, or best practice, as far as maintenance is concerned.

Although no complete service of preventive maintenance was found, several firms conduct annual overhauls of their machinery, in some cases with outside help to make sure that they are done properly. Most plants had some routine to periodically (daily or weekly) lubricate their machines. In a number of cases, managers were not familiar with the concept of preventive maintenance and saw no potential benefit in fixing something that is not yet broken.

One firm has devised an interesting system whereby its equipment is classified in three classes receiving different maintenance attention. Equipment of a critical nature, i.e. whose failure may stop production of the whole plant, such as boilers, compressors, etc., is scheduled for

Table 7.4 Zimbabwe: distribution of sample firms by overall in-house repair and maintenance capability

Category	*Number*	*Proportion (%)*
1. Totally dependent on outside help	3	7.3
2. Self servicing, oiling, replacement of belts or similarly accessible items	9	21.9
3. Same as in 2. plus capacity to strip most machines to replace standard parts or get spares made	20	48.8
4. Complete service of repair and maintenance in house	6	14.6
5. Other	2	4.9
6. Not available	1	2.4

periodic maintenance irrespective of its apparent operating condition. For a second type of machinery, with a potential to seriously disrupt production, its critical components had been identified and sufficient stock of spare parts is always available on hand. Finally, for a third group, besides the periodic oil and greasing done routinely, no special provisions are made.

By sectors, as expected, due to the nature of its activities and equipment, the highest level of repair and maintenance capability is present in the **metalworking** industry, followed by the **food and beverages** industry. Small firms, except those owned and managed by technically inclined individuals, tend to rely more on outside engineering firms while the larger firms tend to have well-established repair and maintenance workshops. Some specialized jobs, like rewinding of electric motors, are sent out in most cases, even in larger firms.

7.5 Industrial engineering

Very little organized activity to continuously assess and improve productivity was found in the firms surveyed. For a number of managers, the productivity concept was not well understood and some took it to mean increases in production. Industrial engineering, as a function, is concerned with the systematic assessment and improvement of plant productivity, but no industrial engineers are employed in the factories visited and there was little evidence, in most of them, of the application of productivity enhancement techniques such as work study, payment systems related to effort, plant layout, etc., or even of sophisticated methods for production planning and inventory control. A number of firms had good manual, and in some case computerized, systems of inventory control but these were generally not integrated with the planning of production. The determination of work standards, where it exists, is based on experience and the task is generally assigned to the plant manager or his foremen. The plant layout is generally assessed only when new equipment is brought in or new construction is under way.

A couple of firms had partially installed the Kawasaki Production System (KPS) which seems to be a fad in some managerial circles in Zimbabwe. The work was done by a local accounting firm with the appropriate Japanese licence. The KPS integrates production, quality control, and productivity techniques, and requires continuous follow up after installation, thus claiming substantial amounts of managerial attention. This all but rules out its application in small firms.[31]

It should be noted that the University of Zimbabwe does not train industrial engineers. The Scientific and Industrial Research and Development Centre (SIRDC) is expected to include a mechanical and production engineering institute and a department of industrial management.

7.6 Safety and environmental control

Concern for safety and, in particular, for pollution, are relatively new topics for manufacturing industries in developing countries. Workplace regulations tend to be quite primitive and worker safety generally does not have a very high priority. Preoccupation for the environment was in the past largely considered a 'rich man's' problem, but is now generating some attention due to the increased incidence of air and water pollution as a result of uncontrolled industrial production.

Responsibility for safety was specifically assigned in 15 cases and for pollution control in five only (see Table 7.5). With seven plants out of 13, **food and beverages** had the highest proportion of plants with safety responsibility assigned.

Written safety rules exist in 15 plants and for pollution control in only one plant. About one-third of the firms carry out periodic *fire evacuation drills*. Half the number of these firms are in the **food and beverages** industry.

Although some firms experienced serious accidents because of the lack of proper machine safeguards and/or worker training, safety con-

Table 7.5 Zimbabwe: responsibility for safety and pollution control in sample firms

Industry	*Safety*		*Pollution control*		*Total*
	Yes	*No*	*Yes*	*No*	
Food	7	6	2	11	13
Textile	2	8	0	10	10
Wood	4	6	1	9	10
Metal	2	6	2	6	8
Total	**15**	**26**	**5**	**36**	**41**

siderations did not seem to be an important concern for a majority of the managers interviewed.

Environmental control received an even lower priority, although some firms had experienced serious pollution problems. Admittedly, the lack of regulations (and/or their enforcement) affects managerial conduct by providing little or no incentive to act in this matter.

8
Technological Functions and Their Organization

The existence of codified information and well-defined technical functions within a plant could be construed as an indication of technological maturity. However, as indicated in prior chapters, size of firm strongly influences the ability to define and to assign these functions within the establishment. In this chapter we examine the existence of technical documentation and technical offices in the plants visited, followed by a review of the practices related to organized technical change at the product and process level, the use of support services, and training.

8.1 General

The existence of technical documentation about process and product requirements, as well as that of technical offices, are examined first followed by an assessment of the overall level of technical organization.

Documentation

The technical documentation used differs among industries, and, for the same industry or type of process, smaller firms are less likely to use technical written documentation than larger firms.

A substantial proportion of the technical know-how utilized in the establishments visited remains uncodified, i.e. has not been spelled out in written instructions. In the **food and beverages** industry, six firms had written recipes, five professed to have manuals and other documentation, one had specifications, and one said that know-how is tacit and diffused by word of mouth via the managing director. In the **textiles and clothing** industry, seven firms had patterns or models, one had blueprints and manuals with the client also providing styles, and two had tacit knowledge diffused by word of mouth via supervisory personnel. In the **wood and furniture** industry, two firms had no documentation, and the other eight used various types of technical documents, including, cutting lists and drawings, sketches, blueprints, instructions and manuals, catalogues, templates, and drawings obtained from customers. Finally, in the **metalworking** industry, one firm had no documentation and the other seven used instructions, blueprints, product specifications, manuals, and drawings.

Technical offices

The questionnaire inquired about the existence of the following technical offices: product engineering, manufacturing or process engineering, quality control laboratory, industrial engineering office, R&D and pilot or experimental plant, plus that of any other technical office, as well as the number of people they employ.

There was a general dearth of technical offices among the firms in the sample, and the most frequently found technical office, but only in six firms (four of which in **food and beverages**) was *quality control laboratory*; it was also the office employing more technical personnel (36 people). Among other offices, *design*, was found in two **clothing** and one **furniture** firms; *drafting* in one **metal-workshop**, and *tool and die-making shop* in two **metalworking** firms; one of these shops was quite sizable employing 10 people.

Level of technological organization

Table 8.1 shows the distribution of firms according to the existence of documentation and technical offices. An index was devised categorizing them in the following way: a **1** was assigned to those firms without documentation or technical offices; a **2** to those with little documentation and/or up to two technical offices, and a **3** to those firms with good technical documentation and/or three or more technical offices. In total, only six firms (14.6%) had a score of **3**, 13 (31.7%) a score of **2**, and 22 or 53.7%, a score of **1**. Interestingly, four of the six better performing firms are in the **food and beverages** industry, but in terms of average performance, the **metalworking** industry does a little bit better (mean score of **2** against **1.77** for food.)

Table 8.1 Zimbabwe: distribution of sample firms by overall index of technological organization

Index	*Frequency*	*Percent*
1. No documentation or files, no technical offices	22	53.7
2. Very little documentation, isolated drawings, charts or sketches; and/or up to two technical offices	13	31.7
3. Good technical documentation and/or three or more technical offices	6	14.6

8.2 Product design and product changes

In practice, it is often difficult to separate product from process changes; nevertheless, for analytical purposes, these two types of technical change are considered separately. Origin of the design of the product, changes made to it and their reason, as well as the introduction of new product designs are considered in this section.

Source

The ability to produce according to own design is more important in such industries as **clothing** than in others as for example, **non-alcoholic beverages**. Nevertheless, some **clothing** firms in our sample were operating as subcontractors that only cut, saw and trim for third parties, which naturally does not entail any design capacity.

According to our respondents, in 44% of the cases the design was self-generated following local specifications, in 30% of the cases it was a copy of a similar imported product, in 14% the design was foreign-specified, in 3% of the cases it was a copy of a similar, domestically produced product, and in 9% of the cases other sources were given. A majority of these cases involved customer-specified designs.

Product changes

A relatively high number of firms (33 or 80%) introduced some changes to the design of the product, but in only five (12.2%) cases were these changes considered substantial in nature by our respondents (see Table 8.2).

The reasons given for the changes in product design given are shown in Table 8.3. A majority of the changes were motivated by the need to

Table 8.2 Zimbabwe: distribution of sample firms by type of change in product design

Change	*Frequency*	*Percent*
1. None	8	19.5
2. Some	28	68.3
3. Substantial	5	12.2

meet new market demands, followed by adaptations to local climate, updating of the design, use of local materials, use of available equipment, and to diversify and simplify the product.

New product introduction

According to our respondents, 20 firms had introduced new products in the last year, a large majority of which were developed in-house and with little or no external help.

In the **food and beverages** industry, help for testing was secured from several external sources. In the **furniture** industry, buyers provided samples or designs, and in a **metalworking** firm, a foreign expert (from Australia) came to assist in product development.

8.3 Technical changes in the production process

Altogether 31 instances of changes were mentioned but some of them do not really qualify as such (e.g. reconditioning of equipment,

Table 8.3 Zimbabwe: reasons stated for product design changes carried out

To	*Frequency*
1. Meet market needs or incorporate new functions	16
2. Suit local tastes	11
3. Update design	9
4. Use of local materials	8
5. Use of available equipment	5
6. Differentiate product	5
7. Simplify the product	4
8. Suit local climate	1
9. Other reasons*	8

Note: *The other reasons were: to suit purchasing power; because new products are being constantly developed; to suit religious restrictions; to satisfy customer; to improve mobility; and to save in raw material (wood).

improvement in the electric power supply). The most frequently mentioned were: capacity stretching (seven cases), energy saving (six cases), adaptation to local materials (four cases), and new tools, dies or fixtures (three cases).

As to performance by industry and size, five firms, *all large*, carried out process changes in the **food and beverages** industry; in **textiles and clothing**, five firms, three large and two medium size, carried out technical process changes; in **wood and furniture**, also five firms, four large size and one medium size, carried out process changes, and in the **metalworking** industry, four firms, three of them large and one medium size carried out such changes.

Thus only large and medium size firms were involved in modifying their production processes.

8.4 Research and development

As indicated above, little applied research activity takes place in the firms interviewed. There are few formally organized R&D offices, and the technical efforts in product development or process changes take place in a few large or medium size firms. Thus, some product development has taken place in the **food** industry to take into account local tastes, income levels, and available raw materials. **Clothing** and **furniture** exporters have had to introduce some modification in their designs to meet specifications from abroad. One **furniture** maker adopted labour-saving hydraulic assembly fixtures, and standardized the product line to attain economies of scale. In **metalworking**, some establishments have the capacity to design simple machinery and equipment for agricultural production, pollution control, etc. Some efforts to adapt to local conditions in electricity supply and to keep old machinery running were also noticeable across industries.

Data on R&D was also collected in a panel survey which included a larger sample (200 firms). Respondents were warned not to consider as R&D routine activities of quality control or maintenance, even if scientists and engineers were employed in carrying them out, and R&D was defined as: 'systematic activity undertaken to increase the stock of knowledge. It includes fundamental research, applied research, and experimental development leading to new devices, products or processes' (UNESCO, 1993).

As can be seen from Table 8.4, the number of firms that declared to be conducting R&D is small, and if one excludes a metalworking

Table 8.4 Zimbabwe: R&D expenditures by sector in panel sample (US dollars)

Sector	*Number of firms doing R&D*	*Amount spent*	*Sales*	*R&D as % of sales*
Food	11	430,143	382,183,288	0.11
Textile	9	70,714	16,689,674	0.42
Wood	18	405,714	249,240,820	0.16
Metal	8	1,527,714	82,701,229	1.85
Total	**46**	**2,434,285**	**730,815,011**	**0.33**

industry outlier, so are the amounts of money spent, total, and as a proportion of sales.

8.5 Use of support services

The support service most commonly used by the manufacturing plants visited was to *repair and maintain* the equipment. This is a reflection of the difficulties experienced by Zimbabwe's industries due to the old age of their machinery and equipment. Thirty-three of the 41 plants visited (80.5%) used some external support service for this purpose (see Table 8.5). Only one of the plants in the **food and beverages** and **textiles and clothing** sectors did not use these services. Two of the plants in **wood and furniture** did not use it either, but one of them is a very small shop producing with hand tools. Most of the plants (four) that could dispense with outside help to repair and maintain their equipment are in the **metal manufacturing** sector, and can do so because the nature of their own production facilities allows them to repair all their equipment in-house. (See Chapter 7, section 4.)

Table 8.5 Zimbabwe: most frequently used support services by firms in sample

Support service	*Frequency*	*Percent*
1. Repair and maintenance	33	80.5
2. Technical assistance	18	43.9
3. Quality control	17	41.5
4. Personnel training	15	36.6
5. Instruments calibration	6	14.6
6. Other	4	9.8

Technical assistance was the second most frequently used support service (18 of 41 plants, or 44%). These services were quite varied, including help to instal new machinery and manufacturing processes, to set up new quality control systems, to train personnel, to design new products, and, frequently, to repair machinery. A peculiar finding in one-third of the 18 cases was the lack of direct payment for the technical assistance services received. In two instances, assistance was provided by shareholders and the parent company at no cost, and in the other four it was supplied by donor governments.[32] Contrasting with repair and maintenance support services, there were very few local suppliers of technical assistance services. Plants in the **metal manufacturing** sector have sought technical assistance more frequently (six out of eight) than those in other sectors.

Quality control support was used by 17 plants (41.5%). Not surprisingly, these plants were clustered in the **food and beverages** sector, where sanitary considerations demand strict quality standards, and in **metal manufacturing**, where some products require strict tolerances. Quality control support services were also used by exporting **clothing** and **furniture** plants.

Training support services were used by 15 plants (36.6%). These were also clustered in **food and beverages** and **metal manufacturing**. Indeed, none of the plants in **textiles and clothing** used them, and only two **furniture** plants did. These services were used mainly to train middle management and technical personnel. Training floor-workers was normally done on the job. In most cases the training was done by local providers and it also took place locally. In a few instances, the suppliers were foreign transnational corporations and equipment suppliers, but even in such cases the training frequently also took place locally.

Instrument calibration support was used only by six firms. Not surprisingly, four of them were in **food and beverages** and the other two in **metal manufacturing**. These services were provided by local suppliers such as the Standards Association of Zimbabwe (SAZ) and the University of Zimbabwe.

Desirable support services

During the survey, interviewees were shown a support service 'wish list' from which to choose and rank desired services. The managers' choices can be seen in Table 8.6. It is noteworthy, first, that they expressed the desire to have support services that differ substantially from those that they are currently using. The most frequently mentioned desired services are to improve productivity techniques (28 firms or 68.3%), personnel training (24 or 58.5%), process improvement (19 or 46.3%), and product design (18 or 43.9%). These are followed by quality control testing (14 or 34.1%), repair and maintenance (13 or 31.7%), trouble-shooting assistance (13 or 31.7%), and energy saving (12 or 29.3%). Assistance with labour relations, pollution control, tools, dies and fixture manufacturing, safety systems, pilot plant experimentation and instrument calibration, were mentioned by less than 20% of the managers and ranked low in their priorities.

Second, their expressed preferences were focused on the need to *increase productivity, change the production processes* and *improve product quality*. These are all new goals for manufacturing plants in the country, and reflect their perceived need to adapt to a changing economic environment. Satisfaction of the managers' wishes would require, besides access to the specific services desired, the introduction of significant changes in their capital equipment and production processes, as well as in the quality of the labour force.

Third, *repair and maintenance, quality control*, and *trouble-shooting* help were second in importance to the assistance mentioned above. The demand for these support services probably reflects the needs of plants

Table 8.6 Zimbabwe: most highly desired support services by firms in sample

Service	*Number interested*	*Number ranked it*	*Average rank*
1. Productivity techniques	28	25	2.00
2. Personnel training	24	22	2.55
3. Process improvement	19	14	2.64
4. Product design	18	13	3.23
5. Quality control testing	14	14	2.50
6. Repair and maintenance	13	11	2.90
7. Troubleshooting assistance	13	11	3.00
8. Energy saving	12	8	2.88
9. Labour relations	8	4	2.80
10. Pollution control	7	5	3.60
11. Tools, dies or fixtures	7	4	2.75
12. Safety	5	4	3.75
13. Pilot plant experimentation	4	1	5.00
14. Instruments calibration	2	0	n.a.
15. Other*	2	2	1.00

Note: *Replies included here are: 'transfer of technological process' and 'to obtain financing for new equipment'.

to continue working with the equipment they have under the assumption that they will not be able to make substantial investments in new plant and equipment.

Fourth, support desires vary across sectors. Assistance with *productivity techniques* was important in all sectors, but it was overwhelmingly important in **food and beverages** where 11 of 13 firms mentioned it and ranked it very high. *Personnel training* was more important for **textiles and clothing** (six of ten firms) and **wood and furniture** (seven of ten firms), the two sectors that are currently using less outside training support. *Product design* was not too important in **food and beverages**, but it was important in the other sectors, specially in **wood and furniture** and **metal manufacturing**. Technical support to improve *safety, pollution control, energy saving* and *instrument calibration* was not desired by any of the **textiles** firms. A majority (four out of seven) of the wishes for *pollution control* support services were expressed by **food and beverages** plants, and most *tools, dies and fixture* manufacture (three out of seven) desires were in **metal manufacturing**. These last results were not unexpected, since pollution problems can have greater impact on food product quality, and tools, dies and fixtures are more frequently required in metal manufacturing.

The panel data offers complementary information to the survey. Table 8.7 shows the number of panel firms using support services from several local suppliers. As one could expect, the largest support supplier was the Standards Association of Zimbabwe (SAZ), which serves mainly **food and beverages** and **metal manufacturing** plants, and was used by a total of 32 of the 200 panel firms. Other suppliers such as the University of Zimbabwe, the Scientific and Industrial Research and Development Centre, the Timber Advisory Board, ZIMTRADE, the Clothing Institute, and the Association of Packaging, served very few panel firms (five or less).

The panel data (Table 8.8) also gives some information about the characteristics of the firms that use support services more frequently. These data show that there were only six firms that had used at least three times either one of the two most frequently used services. One of them was medium size, and the other five were large plants. Four of the six were in **metal manufacturing**. One-half of the plants had foreign ownership.

The above results indicate that: (i) in general, support services are not used extensively by Zimbabwe's manufacturing sector, (ii) as originally hypothesized, the most frequent users tend to be large firms and

Table 8.7 Zimbabwe: number of firms using key support services in panel sample

Sector	*Standards Association of Zimbabwe*	*University of Zimbabwe*	*Scientific & Industrial Research & Development Centre*	*Clothing Institute*	*Timber Advisory Board*	*Zimtrade*	*Association of Packaging*	*Other**
Food	14	3	0	0	0	0	0	1
Textiles	5	0	0	1	0	1	0	6
Wood	3	0	0	0	1	0	0	0
Metal	10	2	2	0	0	0	2	2
Total	**32**	**5**	**2**	**1**	**1**	**1**	**2**	**9**

Note: *Includes: David Whitehead Laboratories, Overseas Institute, Dunlop, Engine Reconditioning Association, Motor Trading Association, and Leather Institute of Zimbabwe. University of Zimbabwe includes: Department of Research and Specialized Services, Department of Metallurgy, and Department of Biotechnology.

Table 8.8 Zimbabwe: characteristics of firms that used support services frequently*

Sector	*Size*			*Ownership*			*Ethnic origin*		
	S	*M*	*L*	*Pr.Z*	*F***	*Sta.****	*Asia*	*Eur.*	*Afr.*
Food	0	0	1	0	1	0	–	–	–
Textiles	0	0	1	1	0	0	–	–	–
Wood	0	0	0	0	0	0	0	0	0
Metal	0	1	3	2	2	0	0	2	0
Total	**0**	**1**	**5**	**3**	**3**	**0**	**0**	**2**	**0**

Notes: * Firms using at least 3 times either one of the two most frequently used services.
** Firms with any kind of foreign participation were classified as foreign.
*** Firms were classified as state-owned when there were 100% in government hands.
– Firms with information not available. In the metal-working industry, two observations lack ethnic origin information.

also those with foreign ownership, and (iii) potential exists for improving the support services available to better match the needs expressed by the reporting firms.

8.6 Training

Workers

In-house arrangements for the training of new workers vary significantly among plants. Paradoxically, a number of the managers interviewed expressed a preference for hiring unexperienced workers. In **food and beverages**, 11 of 13 firms hired such workers, while only four hired formally trained workers or workers with experience in the industry. In **textiles and clothing**, these figures were eight (four and four) out of ten plants. In **wood and furniture**, out of 10 plants, eight hired untrained workers, seven hired formally trained workers, and six hired experienced workers. In **metal manufacturing**, seven of eight plants hired untrained workers, while four hired formally trained, and three hired experienced workers.

In addition to possible interindustry differences in labour skills supplies, these figures seem to reflect managers' expectations about workers' performance. A number of managers expressed serious doubts about the quality of formal vocational training in the country, and thought that students were not just poorly trained, but also mistrained. They felt that many graduates had been taught to do jobs in a particular way, and were reluctant to adapt to the way in which things were done in their plant. Another complaint was that workers who had experience were likely to have also been ill-trained in the previous job. The dissatisfaction with the current training system was particularly strong among managers of **textiles and clothing** firms. Indeed, their producers' association had decided to finance and operate their own school. About 10 years ago they requested government permission to use the education levy on salaries to finance their school. When the government refused, they closed the school down. However, there are now plans to reopen it.

In spite of the expressed preference for hiring untrained, inexperienced personnel, a number of plants did hire trained and experienced people, and in those cases there was a consensus that, on average, trained and experienced workers took significantly less time to reach a satisfactory level of productivity.

The reluctance to hire trained personnel can also have other causes. First, many firms expect their employees to develop a career within the plant. In these plants there is a preference for hiring bright, hard-working young people and groom them within the firm. Several managers expressed the belief that the supply of unskilled labour was so ample that they could choose workers with very high natural ability, who would require relatively short on-the-job training.[33] Second, owner/managers of many mid-size plants began their careers as on-the-floor managers or workers. These people have a strong production background, and little training or experience in personnel management, finance, etc. Since they are likely to want to have things done their way, they would be reluctant to hire trained and experienced workers. Third, most plants have very few technical and supervisory personnel. Most of them have little formal training and are generally white. Indeed, in several occasions the interviewers found that the key quality of the supervisory personnel was their race. In such cases, hiring experienced, or well-trained blacks, would likely create supervisory problems.

Professional and technical personnel

The practice of offering technical and professional personnel training courses was not widespread, except in **food and beverages**. In this sector, eight of 13 plants offered training to their technical personnel, and six plants did the same for professional personnel. Training programmes for professional personnel were least common in **textiles and clothing**, and **wood and furniture**, the sectors that also employed the least number of professionals. In these sectors, and in **metal manufacturing**, technical personnel training was present in about 30% of the plants visited. These programmes included training in new production lines, managerial skills, completing studies for a degree, or graduate university programmes such as the MBA, quality control, and computers. They were provided by local suppliers, parent firms, and universities.

Part IV

Technical Change, Skills, and Firm Performance

9
Planned Technical Change

It was to be expected that the economic policy changes being introduced in the country would have repercussions in terms of new investment projects to improve product quality and plant productivity. Another reason to expect new plant investment was of course the already noted old age of the existing manufacturing equipment. In turn, whether contemplated *a priori*, or not, by the firms planning to acquire new equipment and machinery, investment plans generally also have technical change implications.

The survey included questions about: (a) the existence of investment plans, (b) the nature and expected impact of the technical changes being planned, and (c) the costs of the planned investments.

9.1 Existence of investment plans

Because of the changing economic environment and the very old machinery and equipment in use, positive answers were generally given to the question about the existence of investment plans (see Table 9.1). Only 14, or about one-third, of the plants visited expected to remain unaltered for the foreseeable future. Nevertheless, it was quite difficult to establish how well founded and realistic the investment plans were. A few plants were undergoing drastic changes: updating their equipment, introducing new processes, and expanding their capacity. However, many of the 19 plants that planned to effect changes had not secured the financial resources necessary to carry out the planned changes.

The investment plan responses varied significantly across sectors. **Wood and furniture** and **metal manufacturing** are the industries in

Table 9.1 Zimbabwe: existence of plans for technological development in sample plants

Existence of plans	*Frequency*	*Percent*
1. Remain at same capacity level and with same technology (i.e. no plans for technological development)	14	34.1
2. Remain at same capacity level, but planning to upgrade technology	5	12.2
3. Planning to expand/invest in technological upgrading	13	31.7
4. No definite plans (only ideas, dream)	6	14.6
5. Other*	3	7.3

Note: *Answers included here are: 'to relocate in a different city', 'to shrink', and 'to have a larger plant with similar technology'.

which more technological change is foreseen. In each of these sectors, six out of ten firms want to upgrade technology. In **wood and furniture**, five of them want to expand output, and in **metal manufacturing** three plants want to do the same. In **metal manufacturing**, only one plant expects to remain unchanged. **Textiles and clothing** present a very mixed picture. Three plants have no plans, one wants to upgrade technology and expand, and five fall in the 'other' category that includes those that wish to expand output with the same technology, those that wish to shrink, and those that want to move to another city. The largest proportion of plants that do not expect changes was found in **food and beverages** where seven (54%) fell in that category. Five plants want to upgrade technology; four of them are also planning to expand.

9.2 The nature and expected impact of technical changes planned

Furniture exports are growing and firms from this industry in the sample expressed the need for: new equipment (seven out of ten plants), to hire technical personnel (four plants), and to contract a licence (four plants). The goals of the planned technical changes in this sector were: to improve the quality of the products and to lower costs (six plants each), to introduce new products, and to increase productivity (five plants each), to introduce new processes (four plants), and to subcontract (two firms).

The **metal manufacturing** plants visited also plan to make significant changes. Six out of eight plants expressed a need for new equipment and three plants for technical assistance agreements. These changes were expected to result in: lower costs and increased productivity (four firms each), new products, increased product quality and new processes (three plants each), and to involve some subcontracting (two plants).

In **food and beverages**, only three plants want new equipment. They expected that their plans would result in improved product quality, new processes, higher productivity, and lower costs, in four cases each, and in the introduction of new products in three plants. One plant expected to do some subcontracting.

In the case of **textiles and clothing**, only three plants want new equipment. The planned changes are expected mainly to increase productivity (three plants), and lower costs (two plants). One plant each wanted to introduce new products, improve existing product quality, and engage in some subcontracting.

To implement their plans, plants (mainly those in **food and beverages** and **metal manufacturing**) will have to train existing personnel. These plans also call for external assistance to de-bug the new equipment, mainly in **wood and furniture** and **metal manufacturing**. The equipment changes would also require organizational changes, but only in a few firms (two each in **textiles and clothing**, and **wood and furniture**, and one in **metal manufacturing**). None of the plans of **food and beverages** firms contemplates organizational changes.[34]

The planned investments are expected to improve the competitive position of all **food and beverages** and **wood and furniture** plants involved, and of four out of six **metal manufacturing** plants that plan investments. In the **textiles and clothing** sector the picture is quite different, as only two plants expect to achieve this kind of result.

9.3 The costs of planned investments

One of the main signs of the uncertainty surrounding these investment plans is the scarcity of cost estimates (see Table 9.2). In the **food and beverages** industry, three of the plants that wanted changes had estimated the cost of the desired investments. Only two **textiles and clothing** firms had estimates, and they were quite low ($8,500 and $37,000). In these two sectors, investment plans are not likely to

Table 9.2 Zimbabwe: estimated cost of planned technological changes in sample

Firm (by sector)	*Employment*	*Ownership*	*Ethnic origin*	*Total cost estimate (US$)*
Food				**$ 457,143**
1351	4	Private(Z)	Asian	not appl
1731	21	Private(Z)	European	$ 28,571
551	49	Private(Z)	African	not appl
211	54	Private(Z)	European	not appl
11	126	St&Priv(F)	European	not avail
591	150	Private(F)	European	$ 214,286
1661	312	Private(Z)	European	not appl
1951	312	St&Priv(F)	not avail	not appl
1401	360	Private(Z&F)	not avail	not appl
241	364	St&Priv(Z)	European	$ 214,286
361	799	St&Priv(F)	European	not avail
1221	839	State	not appl	not appl
71	1131	Private(Z)	Other*	not avail
Textiles				**$ 45,814**
161	2	Private(Z)	African	not appl
751	11	Private(Z)	Asian	not appl
461	12	Private(Z)	African	not appl
1121	47	Private(Z)	European	not appl
1681	49	Private(Z)	European	not appl
1701	95	Private(Z)	African	not appl
1581	156	Private(Z)	European	not appl
681	230	Private(Z&F)	not avail	$ 8,571
51	450	Private(Z&F)	European	$ 37,143
271	1554	Private(Z)	European	not appl
Wood				**$1,706,572**
41	4	Private(Z)	African	not appl
131	14	Private(Z)	European	not appl
1471	77	Private(Z)	European	$ 314,286
1771	107	Private(Z)	African	$ 428,571
1191	115	Private(Z)	not avail	$ 175,143
331	195	Private(Z)	European	not appl
231	197	Private(Z)	Mixed	$ 71,429
1861	220	Private(Z)	European	not avail
471	236	Private(Z)	European	$ 717,143
1921	366	Private(Z)	not avail	not avail
Metal				**$3,444,284**
521	24	Private(Z)	Mixed	$ 11,428
1361	43	Private(Z)	European	not avail
1751	75	Private(Z)	Mixed	not appl
1141	88	Private(F)	European	$1,285,714
1641	96	Private(Z)	European	$ 690,000
581	150	State	not appl	not appl
491	528	St&Priv(F)	European	$1,357,143
411	1353	Private(Z)	not avail	$ 100,000
Total				**$5,653,813**

Note: *A public corporation.

produce significant changes, at least in the short and medium run. There is a greater number of plants with planned investment cost estimates in the **wood and furniture** and **metal manufacturing** industries. Five plants in **wood and furniture** estimated planned investments with costs ranging from $71,000 to $717,000. In **metal manufacturing** two plants had modest plans for investments of $11,400 and $100,000, but those for three other plants ranged from $690,000 to $1,357,000.

All in all, the answers about investment plans provide a picture of firms that, in different ways, are generally trying to respond to the challenges posed by the new economic policies adopted by the government by improving their manufacturing plants. This attitude is more clearly evident in the **wood and furniture** and **metalworking industries**.

10
Skills and Technological Capability

Firm technological capabilities and performance will depend, in an important measure, on the presence of professional and middle-level technical skills. Such skills seem to be quite sparsely distributed in Zimbabwe's manufacturing. In this chapter we review the results of survey questions about the existence of high-level technical skills in the firms and industries studied. The possible correlation between indicators of firm technological competence and economic performance is also explored.

10.1 High-level technical skills

The information available is from the firms interviewed as well as from the panel sample, classified by sector. Comparison is also possible with data from another Sub-Saharan Africa country.

Sector and country data

As can be seen from Table 10.1, the proportion of *professional and technical personnel* is quite low in the firms interviewed, particularly so for the first type of personnel. With a few exceptions, there is an association between size of firm, as measured by employment, and the number of professional and technical personnel. In a couple of cases, one in the **food and beverages** industry and the other in **wood and furniture**, the number of technical personnel seems to be overstated.

A comparison with a similar sample for Kenya (see Table 10.2) shows that while in **textiles and clothing** and **wood and furniture** the proportions of professional personnel are quite similar, Kenya clearly surpasses Zimbabwe in **food and beverages**, where it

Table 10.1 Zimbabwe: professional and technical personnel in sample

		*Professional personnel**		*Technical personnel***	
Firm (by sector)	*Employment*	*Number*	*%*	*Number*	*%*
Food	**4521**	**51**	**1.13**	**324**	**7.17**
1351	4	1	25.0	0	0.0
1731	21	0	0.0	2	9.52
551	49	1	2.04	2	4.08
211	54	0	0.0	6	11.11
11	126	0	0.0	4	3.17
591	150	n.a.	n.a.	5	3.33
1661	312	1	0.32	3	0.96
1951	312	6	1.92	8	2.56
1401	360	4	1.11	3	0.83
241	364	1	0.27	2	0.55
361	799	13	1.63	16	2.0
1221	839	9	1.07	256	30.51
71	1131	15	1.33	17	1.50
Textiles	**2606**	**22**	**0.84**	**34**	**1.30**
161	2	0	0.0	1	50.0
751	11	0	0.0	1	9.09
461	12	0	0.0	7	58.33
1121	47	1	2.13	1	2.13
1681	49	0	0.0	2	4.08
1701	95	0	0.0	2	2.11
1581	156	1	0.64	1	0.64
681	230	1	0.43	8	3.48
51	450	2	0.44	2	0.11
271	1554	17	1.09	9	0.58
Wood	**1531**	**4**	**0.26**	**190**	**12.41**
41	4	0	0.0	4	100.0
131	14	0	0.0	0	0.0
1471	77	0	0.0	1	1.3
1771	107	1	0.93	23	21.5
1191	115	0	0.0	15	13.04
331	195	1	0.51	129	66.15
231	197	1	0.51	7	3.55
1861	220	0	0.0	5	2.27
471	236	0	0.0	1	0.42
1921	366	1	0.27	5	1.37
Metal	**2261**	**28**	**1.24**	**149**	**6.59**
521	24	2	8.33	5	0.21
1361	43	1	2.32	6	13.95
1751	75	0	0.0	6	8.0
1141	88	1	1.14	7	7.95
1641	96	0	0.0	2	2.08
581	150	5	3.33	6	4.0
491	528	6	1.14	43	8.14
411	1351	13	0.96	74	5.48
Total	**10919**	**105**	**0.96**	**697**	**6.38**

Notes: *With university degree
**With certificate or diploma.

Table 10.2 Comparison of the proportions of professional and technical personnel by sector in Zimbabwe and Kenya samples

Sector	*Professional personnel (%)*		*Technical personnel (%)*	
	Zimbabwe	*Kenya*	*Zimbabwe*	*Kenya*
Food	1.13	3.42	7.17	5.23
Textiles	0.84	0.85	1.30	3.14
Wood	0.26	0.00	12.41	1.17
Metalworking	1.24	2.44	6.59	6.82
Total	**0.96**	**1.64**	**6.38**	**4.43**

has triple the proportion of professional personnel, and in **metalworking**, where it has almost double the proportion of such personnel.

With respect to *technical personnel*, as shown in the same table, Zimbabwe clearly has a higher proportion than Kenya in **food and beverages**, and a much higher one (more than 10 times greater) in **wood and furniture**, while Kenya has a substantially higher (two-and-a-half times) proportion in **textiles and clothing**, and a somewhat higher one in **metalworking**.

For the total sample, Kenya has 1.64 professionals per 100 employees and Zimbabwe 0.96, while in technical personnel, Zimbabwe has 6.38 per 100 employees and Kenya 4.43. Clearly, the deficit of high-level skilled personnel, especially engineers, observed in Zimbabwe's manufacturing plants, is greater even than in Kenya.

Comparison of the data for larger samples in both countries conveys a similar picture, particularly with respect to engineers (see Tables 10.3 and 10.4). Kenya has a higher proportion of

Table 10.3 Zimbabwe: employment of scientists, engineers and technicians by sector in panel sample

Sector	*Total employment*	*Scientists*		*Engineers*		*Technicians*	
		Number	*%*	*Number*	*%*	*Number*	*%*
Food	15,983	103	0.64	22	0.14	280	1.75
Textiles	2,496	0	0.00	3	0.12	58	2.32
Wood	32,411	1	0.00	38	0.12	353	1.09
Metal	5,479	1	0.02	41	0.75	115	2.10
Total	**56,369**	**105**	**0.19**	**104**	**0.19**	**806**	**1.43**

Table 10.4 Comparison of the proportions of scientists, engineers and technicians by sector in Kenya and Zimbabwe panel samples

Sector	*Scientists (%)*		*Engineers (%)*		*Technicians (%)*	
	Zimbabwe	*Kenya*	*Zimbabwe*	*Kenya*	*Zimbabwe*	*Kenya*
Food	0.64	0.34	0.14	0.29	1.75	2.54
Textile	0.0	0.05	0.12	0.35	2.32	0.90
Wood	0.0	0.10	0.12	0.21	1.09	0.82
Metal	0.02	0.12	0.75	2.45	2.10	6.33
Total	**0.19**	**0.15**	**0.19**	**0.74**	**1.43**	**2.42**

engineers per 100 workers in all four sectors. Scientists, most likely a majority of them chemists, are more prevalent in the **food and beverages** industry due to chemical testing required for quality control. Engineers are more concentrated in **metalworking** in both countries. For the total samples, the proportions of scientists are not very different between the two countries, but in engineers, Kenya has almost four times the proportion than Zimbabwe and with respect to technicians about 70% more.

Thus high-level technical personnel is very scant in Zimbabwe's manufacturing, in spite of the country's post-independence educational achievements and its higher level of industrial development relative to other Sub-Saharan Africa countries.

Firm data

Although substantial variability obtains within each sector, a large majority of professional personnel, as well as technicians, are concentrated in the largest firms, a number of which are foreign, state-owned, or joint ventures. Among firms in the panel, about 48% of the total number of engineers and about 40% of the technicians are employed by the three largest employers (see Table 10.5). Also, as mentioned above, the sample data also shows clear correlation between employment of professional and technical personnel and size of firm (see Table 10.1).

10.2 Technological competence and firm economic performance

While it might be reasonable to assume that the level of economic performance would be associated with technological capability, it is not

Table 10.5 Zimbabwe: concentration in employment of scientists, engineers and technicians (panel sample)

Personnel	*Total number*	*Number of employing firms*	*Number employed by 3 largest employers*	*% of total number*
Scientists	105	8	1	0.95
Engineers	104	21	50	48.08
Technicians	806	52	320	39.70

easy, first, to adequately measure technological competence, and second, to try to relate it to indicators of economic efficiency. An attempt is thus made in this section to create an aggregate measure of technological capability, and then to study its correlation with various indicators of economic performance. In turn, and as stated in Chapter 5, section 3, technological capability was expected to be associated with size of firm and foreign connections.

Survey data

Data on three possible indicators of economic performance: rate of profits, sales per worker and proportion of output exported, were examined for the sample firms. Each of these indicators poses problems to the analyst. The profits data seem unreliable. Sales per worker, as a rough productivity indicator, has the problem that, even for the same industry, differences in product mix (and thus in the proportion of value added), and in capital–labour ratios, are so large as to make the numbers not directly comparable. Although dependent on export incentives provided, the proportion of output exported is probably the most robust indicator of economic performance of the three. However, so far few firms export a substantial proportion of their output, and some may be pricing their exports at variable cost. The mean value of the proportion of output exported is only around 10% for **food, textiles** and **metalworking**, and around 22% for **wood and furniture** which has the best export performance. The large exporters, one in textiles and clothing and three in wood and furniture, are firms with relatively large size and higher number of technical personnel, thus confirming our prior hypotheses.

Panel data

Taking advantage of the larger size of the panel sample, the relationship between the technology data and key hypothesized explanatory variables, as well as with some firm economic performance indicators, was examined statistically. Two different statistical analyses were attempted: the first, trying to relate technology to size and ownership, and the second, using technology as a variable, among others, to account for economic performance. The possible incidence of sectoral effects was also taken into consideration.

(a) Technology variable

The first task was to construct a technology variable. The completed panel (larger sample) questionnaires contained information on **formal channels of technology acquisition, technical skills, and technological efforts**. The questions pertaining to the formal transfer of technology included: number of technology licences held, number of technical assistance contracts, and number of expatriate personnel. With respect to technical skills, information was requested on: number of scientists, engineers, and technicians employed. Finally, the technical efforts questions dealt with: amount of money spent in R&D, number of scientists, engineers and technicians employed in R&D, and number of uses made of technical support services. An aggregate, synthetic, technology variable was thus constructed including information on the three components: **transfer, skills and effort, mentioned above**.[35]

(b) Technology, size of firm and foreign ownership

The first statistical exercise consisted of trying to estimate the statistical association between the technology variable and two potential 'explanatory' variables: foreign ownership and size of firm. It was hypothesized above (see Chapter 5) that firms with foreign ownership would have facilitated access to technical information; thus, *ceteris paribus*, they should be expected to score higher in the technology variable. Size of firm was also hypothesized to be positively associated with technology access due to the greater division of labour and ability to manage high-level technical personnel present in large firms. As also stated in Chapter 5, for reasons of mandated quality control and required precision, respectively, a higher level of technological activity (or capability) was expected to be found in **food and beverages** and **metalworking** than in other sectors.

Technology variable to be 'explained'

$$T = Tt + Tsk + Teff \tag{1}$$

where

Tt = number of licences + number of technical assistance contracts + number of expatriates.
Tsk = number of engineers + number of scientists + 0.5 number of technicians.
Teff = money spent in R&D (normalized) + number of uses of technical support services (normalized).

'Explanatory' variables

F = 1 if proportion of foreign ownership > 0, otherwise 0.
S = Firm employment, normalized.
X1 = 1 if firm in food industry, 0 otherwise.
X4 = 1 if firm in metalworking industry, 0 otherwise.

Model

$$T = a + bF + cS + \epsilon \tag{2}$$

where

b > or = 0; and c > or = 0

Results of the OLS regression of ***T*** as a function of ***F*** and ***S***, in logarithmic form, are shown in Table 10.6.

As can be seen, for model I, the coefficients of the 'explanatory' variables are of the hypothesized sign and statistically significant, at least at the 0.05 level. For model II, with the dummy variables for the sectors hypothesized to have stronger technological inputs in their activities (**food and beverages** and **metalworking**), the significance of the coefficient of foreign ownership decreases while that for size

Table 10.6 Zimbabwe: technology as a function of foreign ownership, size and industry (panel sample)

Model	*Coefficients and t statistics (n = 198)*							
	Cons.	*F*	*S*	*X1*	*X4*	*Adj R2*	*F stat*	*D.W.*
I	**0.750**	**0.169**	**1.365**			**0.44**	**77.82**	**1.78**
	1.568	1.972	10.35					
II	**1.129**	**0.150**	**1.406**	**−0.10**	**0.171**	**0.45**	**42.16**	**1.86**
	1.442	1.747	10.76	−1.35	2.039			

Note: Parameter estimates are in bold; numbers below are the **ts**.

(employment) increases slightly. The proxy variable for the **food and beverages** industry is of the wrong sign and not significant, while that for **metalworking** is significant and comparable in size to the coefficient of foreign ownership. Thus the **food and beverages** firms in the Zimbabwe panel do not show the technical characteristics hypothesized in Chapter 5, while **metalworking** firms behave as expected. Of course, the results of model II could have been affected by multicollinearity.

It should be noted that, as estimated in both models, the size coefficient is much more important and significant than the foreign ownership coefficient. The size coefficient could be interpreted as the elasticity of 'technology' with respect to scale, and the estimated value of about 1.4 indicates that a 10% increase in establishment size results in a 14% increase in technology attributes (transfer, skills, and technical effort).

(c) Technology in a production function

The second part in the statistical analysis dealt with the relationship between the technology variable and indicators of firm economic performance. First, an attempt was made to fit a simple production function, with and without the technology variable. The model used is shown below.

$$\boldsymbol{Q = a + bK + cL + dT + \mu} \qquad (3)$$

where

$\boldsymbol{Q}$ = sales
$\boldsymbol{K}$ = replacement value of capital assets
$\boldsymbol{L}$ = employment
$\boldsymbol{T}$ = technology variable.

where: b, c, and d > or = 0

OLS logarithmic regression results are shown in Table 10.7.

While the fit is quite good, substantial multicollinearity obtains between the key 'explanatory' variables, labour and capital. Collinear variation between employment and the technology variable is also to be expected on account of the previously shown association between employment size and technological attributes (see b).

As can be seen from Table 10.7 and equation (3), all coefficients are of the right sign but while the capital and labour coefficients are

Table 10.7 Zimbabwe: manufacturing Cobb–Douglas production function

Model	*Coefficients and t statistics, n = 193*						
	Const.	*K*	*L*	*T*	*Adj R2*	*F*	*D.W.*
I	**3.635**	**0.506**	**0.841**	**0.047**	**0.876**	**457.09**	**1.47**
	6.19	9.15	8.74	1.62			
II	**3.053**	**0.530**	**0.876**		**0.875**	**678.53**	**1.49**
	6.53	9.90	9.29				
III	**8.139**		**1.506**	**0.118**	**0.823**	**449.23**	**1.56**
	21.2		19.9	3.51			
IV	**2.146**	**0.871**		**0.103**	**0.827**	**464.06**	**1.50**
	3.23	20.3		3.08			

Note: Parameter estimates are in bold; **t** statistics are the numbers below.

highly significant, that corresponding to technology is the smallest and only significant at the 0.10% level. Dropping the capital variable reduces the proportion of accounted for variance from 87.6% to 82.3%, that is by 6%, while elimination of the labour variable decreases the adjusted R2 to 0.827, for a net reduction of the explanatory power of 5.6%. Finally, dropping the technology variable has minimal impact, reducing the R2 only by 0.001 for a total reduction in explanatory power of 0.1%. From the estimated production function could be read the output (sales) elasticity with respect to capital: 0.51; to labour: 0.84 (which indicates increasing returns to the physical factors); and with respect to technology: 0.05. Of course, given the extent of multicollinearity present, the estimates of the coefficients and the values of the **ts** may be overstating the strength of the existing associations.

(d) Technology and firm performance indicators

Various possible indicators of firm performance were selected initially: sales per worker, value added per worker, and rate of profits. Data to calculate value added was unreliable. It was tried to compute it in two different ways, one by deducting the value of purchases from total sales, and the other by adding up profits declared and wages paid. Profits declared data seemed to be particularly unreliable. Sales per worker was finally selected as the less unreliable among the above mentioned indicators. A second indicator chosen was whether the firm exported, without attention to the proportion of output exported since exports of manufactures are still an incipient phenomenon. Thus, firms were classified in exporting or non-exporting. It was postulated that the performance indicators will be a function of: size (possible economies of scale), capital intensity, and technology. Size was conventionally measured by number of workers employed. For the capital intensity indicator two different measures of capital were tried: book value, and replacement value. The later was adopted because of better availability of data in the sample. The data set had to be 'cleaned' before attempting any statistical analysis because a number of observations had recorded either zero sales or zero employment; in both cases, such observations were eliminated from the sample.

Variables to be 'explained':

Ql : Sales per worker
X : Equal to 1 if exporting firm, and to 0 otherwise.

'Explanatory' variables:

Kl : Capital–labour ratio
L : Employment
T : Technology

Model

$$Ql = a + bkl + cL + dT + x \quad (4)$$

$$X = a' + b'Kl + c'L + d'T + x' \quad (5)$$

where it is hypothesized that: *b*, *b'*, *c*, *c'*, *d*, and *d'* are > = 0. Results of the OLS regressions in logarithmic form are shown in Table 10.8.

Table 10.8 Zimbabwe: firm performance indicators as a function of capital intensity, size and technology

Model	*Coefficients and t statistics, n = 193*						
	Const.	*Kl*	*L*	*T*	*Adj. R2*	*F*	*D.W.*
I (*Ql*)	**–5.525**	**0.506**	**0.348**	**0.047**	**0.614**	**103.44**	**1.47**
	–9.23	9.150	5.311	1.622			
II (*X*)	**0.899**	**–0.267**	**0.883**	**0.283**	**0.428**	**49.198**	**2.01**
	0.684	–2.199	6.146	4.427			

Note: Parameter estimates in bold. Numbers below are the **ts**.

For the sales per worker indicator, the coefficients are all significant and of the right sign. A little over 61% of the variance is explained by this model (I). The technology variable is only significant at the 0.10 level. For the exports indicator, the share of explained variance is smaller (43%), the technology coefficient is larger and highly significant, but the capital–labour ratio coefficient is negative and significant. Of course, the use of OLS regression for a discrete variable such as the exports indicator could be a source of estimation error.

It could be concluded that the panel data shows a statistically significant, though not very strong, effect of the technology variable on both output and output per worker. This effect seems to be stronger in determining whether firms export or not, than in affecting productivity (output per worker).

Part V

Conclusions and Policy Interventions

11
Conclusions

In this chapter, the conclusions of the study are briefly summarized following the order in which they were developed in the book.

11.1 Country background

The country's recent political and economic history has conditioned the evolution of its manufacturing sector and the technological capabilities within it. The autarky prevailing during the UDI period and subsequent economic blockade had the effect of challenging the technical ingenuity and entrepreneurial motivation of its industrialists, but lack of competition and interaction with the outside world resulted in an overgrown manufacturing sector with little concern for quality or innovation and an overriding concern for keeping the machinery running, practically at all costs. New economic policies that include liberalization and opening up to international trade represent a new and serious challenge for the existing industrial firms.

The country is well endowed with agricultural and mineral resources, but is deficient in terms of modern infrastructure in the areas of water and power supply (very exposed during periodic droughts), transport, and communications.

Present social problems derive from its colonial background and the still prevailing racial divide. Although the polity is now in Black hands, the tiny (less than 1% of the total population) White minority controls a great deal of the economic activity, particularly in the private sector. Access of blacks to entrepreneurship has been limited by the discrimination of the past and the present lack of assets and financial opportunities to start new ventures. Tribal animosities and political corruption are also factors negatively affecting industrial and technological development.

11.2 Manufacturing industry

As stated above, industrialization was fostered during periods of forced protection and oriented towards substituting imports for the domestic market. Lack of competition resulted in little or no concern for quality or productivity improvement and little effort was made to promote exports. Nevertheless, the country's large and variegated manufacturing sector managed to achieve some industrial exports essentially due to its low labour costs.

The country has a large supply of unskilled labour due to high rates of demographic growth, but it has also improved its education, specially after attaining independence in 1980. The capital stock is quite old and in need of maintenance and renewal. It also embodies older, simple technologies, which in most cases require updating or change.

Mining represents a great potential for further processing and addition of value. Other local natural resources of possible use in industry are quite limited.

Most manufacturing establishments are concentrated around the capital city of Harare, and in Bulawayo. Large firms are mostly in foreign hands or are para-statal.

11.3 Education, science and technology

The country carried out a very significant educational programme since independence which has resulted in a high degree of adult literacy and rates of high-school enrolment comparable to the best among developing countries, and far ahead from most other Sub-Saharan Africa countries.

It lags however in terms of higher and technical education, in particular as it refers to the production of engineers and scientists able to work in industry.

There is very little research and development activity, and the results of such work in terms of patents and publications are quite meagre, though not below African regional standards.

11.4 Acquisition of technology

Most of the plants visited were set up without the benefit of prior formal feasibility studies. Their equipment was mostly imported and in a large proportion obtained second-hand. A substantial share of this equipment is of old age.

In many instances they received external support for the initial installation of the machinery, but the use of disembodied technology, in the form of licences or technical assistance contracts, was very limited.

11.5 Technical operating capabilities

Local entrepreneurs have acquired a good capacity to troubleshoot and keep running their equipment, which as noted above is generally of an old age.

Quality control of the final product was carried out in most plants but mostly relying on visual and other practical, non-sophisticated, methods. Raw materials and in-process quality control was much more limited, with food industries being generally an exception due to health regulations. Statistical quality control methods are not being used.

Maintenance and repair was a widely spread and necessary technical capability to keep the used and old equipment running. This capacity, as well as trouble-shooting mentioned above, were a necessity during the time of the blockade when the country was forced to use its own resources and ingenuity to keep the equipment going. Nevertheless, there was little in the way of programmed maintenance, and very limited application of preventive maintenance.

The use of industrial engineering to improve productivity was practically non-existent, although a couple of firms were experimenting with new Japanese methods of production management and plant organization.

Environmental control had a very low priority in most plants, as indicated by the lack of programmes and devices to foster it except for some plants that were experimenting serious pollution difficulties. Safety programmes were non-existent, or rudimentary, and few machines had guards or protections. In part this could be attributed to the age of the equipment, since more modern machinery tends to include anti-pollution and safety protections.

Thus, on the whole, existing plant technical capabilities mostly entailed trouble-shooting and repair skills needed to keep the equipment running. There also was some concern for the quality of the final product, but little or no use of technically-based quality control methods.

11.6 Technological functions and their organization

Differentiated technical functions were only present in the largest establishments. Again, except for the large firms, there was a very limited presence of formal technical documentation, and little or no organized research and development activity.

The use of external support services was quite sparse, although the need for such services was generally recognized by management.

There was limited organized internal training of shop-floor workers, with some use of programmes external to the firm for the training of other personnel.

11.7 Planned technical change

Although faced with tariff reductions and the prospect of increased competition, there was little in the way of planned new plant investments embodying technical change or new technologies in the firms surveyed. Uncertainty surrounding the future of the new economic policies might be a cause of this paucity of investment planning.

11.8 Skills and technological capability

As hypothesized, high level technical skills were mostly present in the large firms, and size of firm was shown to be more closely associated with the presence of such skills than foreign ownership.

Among the sectors, **food and beverages** and **metalworking** have higher proportions of such personnel than the other sectors.

A positive, though not very strong, statistical relation was verified between an aggregate measure of technology and some indicators of economic performance at the firm level.

12
Policy Interventions

The justification for intervention to improve economic efficiency in market economies is generally predicated on the existence of market failures due to externalities, economies of scale, poorly, or non-functioning, markets, missing factors, etc.[36]

When dealing with technology, two main sources for such failures can be distinguished. At the 'product' level, technology exhibits substantial public good characteristics, such as limited appropriability as well as uncertainty and irreversibility in the investment processes required for its acquisition, operation and creation. At the 'user' level, in addition to fractured markets for many factors and services, small firms face particular problems due to their lack of high-level skills, technical offices, and the ability to utilize external technical services.

Still, as experience has shown, it is not enough to find grounds for intervention that are in principle acceptable, it is also necessary to justify it with a strong presumption that intervention will lead to an overall improvement, and not to a further deterioration of the relevant economic environment. Of course, conditions vary in different countries, but there is some experience from which to learn as to lesser or stronger grounds for intervention, as well as with respect to the type of interventions more likely to succeed.

Industrial technology policies in Zimbabwe must be considered within the special context presently faced by the country. First, overall economic policies have recently experienced drastic changes. Entrepreneurs, who had mastered ways to operate successfully under the old rules are now learning to adapt to new, quite different, rules. Moreover, recent economic policy changes may not be perceived in all quarters as permanent, and this of course influences entrepreneurial responses to the new policies. Thus, current responses, including their technological components, should not be interpreted as fully reflecting the potential ability to adapt of manufacturing entrepreneurs.

Second, the external (to the country) economic environment has also been changing drastically and is likely to continue doing so. International changes currently taking place beyond Africa, but also in South and Central Africa countries, will undoubtedly have substantial effects on Zimbabwe's manufacturing industry. These potential changes add significant uncertainty to the economic environment in which Zimbabwe's industry is presently operating.

Third, there is an urgent need to promote black entrepreneurial and managerial skills. This should be done not just because of reasons of greater efficiency and equality, but more important, because the current combination of a polity controlled by blacks and an economy largely controlled by whites, if left unchanged, may result in political and social instability.

There is ample scope in Zimbabwe for industrial technology policies. In the formulation of such policies care must be taken to consider the need for coordination with related policy areas such as international trade, education, and science and technology. Selected areas to be considered for potential intervention are discussed below.

12.1 Industrial policy

As mentioned above, in modern manufacturing, industrial, human resources, and trade policy need to be considered in a coordinated fashion. Industrial policy encompasses a number of issues at the national and sectorial levels. Support services, entrepreneurship, and infrastructure, are the key issues herewith considered.

Support services

The manufacturers surveyed make little use of support services, and when they do, it is mainly to repair and maintain their equipment. It is also common for firms, even large ones, to expect not to pay for some of the support services they receive as they have become accustomed to donor countries, NGOs, and holding companies providing these services free of charge. The changes that are taking place in the environment in which Zimbabwe's manufacturing sector has to operate will require a large increase in the use of support services. This will be particularly true for many small and medium size plants that will be competing with large plants that can obtain some of those services from within the firm.

As shown in Chapter 8, section 5, besides support to maintain and repair machines and equipment, the manufacturing sector will require support services to improve productivity and to control and improve product quality, to modernize its equipment, to improve and continuously change product design, to identify, implement, and adapt, new technologies, to improve its marketing techniques to search and conquer new markets, and so on. These changes will require the development of a support service network that was not necessary under UDI and the closed economy of the 1980s, and that exists in all successfully industrializing, and industrialized, countries.

Industrial support services generate significant external economies, and have some public good characteristics that justify government intervention. In many cases, support services supply knowledge, a public good, that under welfare maximizing conditions should be transmitted to the buying firms at marginal cost. A large industrial base is required for a private market to develop, and support, local suppliers of such services. Given the size of its industrial sector, this has not yet happened in Zimbabwe and is unlikely to take place, of its own, in the near future. Moreover, in some instances, as when technologies are first brought into a plant and adapted, the need for support services is only temporary in nature, and not foreseen to grow very much when the prospects for plant investment remain rather sluggish (see Chapter 9).

Entrepreneurship and investment behaviour

Again, one could ask why intervene? Basically, entrepreneurs cannot be made and the distribution of entrepreneurial talent need not be different in Zimbabwe from elsewhere. Thus it could be argued that, in due time, entrepreneurs will arise. While there is some truth in this line of reasoning, experience shows that the conditions faced by local entrepreneurs in developing countries, even when they have the same abilities and talent as those in advanced economies, are quite different and little conducive to allowing their talent and abilities to emerge. Moreover, the prolonged absence of competition may have stifled the development of potential entrepreneurial capabilities. It is also the case that even in some of the most advanced of the industrializing countries, a number of industrial projects had to be launched, at least initially, by the government, because private entrepreneurs did not have the investment capital or could not appropriately discount the long-term benefits accruing from certain undertakings.[37]

Thus although it is largely true that entrepreneurs cannot be formed, it is possible to help their development by removing obstacles and finding mechanisms that induce their emergence. In many cases, access to credit acts as a definite constraint because starting businesses do not have the collateral assets, or even the personal references, necessary to meet banking credit conditions. The availability of venture capital is a problem in most countries, including some industrialized countries, and measures could be taken by the government to create a venture capital window as part of programmes to support the development of small and medium size enterprises.

The reality is that entrepreneurial skills in Zimbabwe are undeveloped and misused. First, most entrepreneurs are white, and frequently feel threatened by the post-independence political situation. In spite of several expressions during the interviews of their desire to leave the country, as noted above, it is unlikely that many will, and it is probably true that those more prone to do so have already left. However, those who stayed have not been encouraged to invest and to contribute significantly to economic growth. Second, social stability and simple fairness requires that blacks have opportunities to develop their entrepreneurial abilities. Financial market segmentation and lack of trust across races are strong obstacles to the expansion and strengthening of black entrepreneurship.

Although for different reasons, both black and white entrepreneurs in Zimbabwe tend to perceive future investments as entailing very high costs. It is true that the access to formal and informal capital markets is more restricted for blacks than for whites who possess greater collateral, wealthier families, and well-established networks of friends. For many blacks the difficulty they experience in having access to credit makes them consider the cost of future investments as possible only at very high interest rates. In the case of whites, in spite of their greater access to credit, the risk associated with foreseen threats of social and political instability has the same effect. The political rhetoric of the past and the diversity of legislation toward foreign investment also result in a similar attitude on the part of potential foreign investors. In sum, although for different reasons, all entrepreneurs in Zimbabwe tend to consider future investments as being possible only at relatively very high real rates of interest.

In order to achieve the investment levels required to increase the rate of economic growth, the country needs to lower the discount rate implicit in the economic behaviour of its entrepreneurs. To achieve this it is necessary to develop credible policies that promote long-run social, political and economic stability, and to develop financial systems and insurance mechanisms that allow the emergence of risk sharing, contingency contracts, and the generation of a supply of credit and venture capital to which all Zimbabweans could have access at reasonable terms. The investment needed for growth, and the increased black participation in management and entrepreneurship necessary for efficiency, social stability, and justice, will not materialize unless barriers to access and the real cost of investment credit are lowered. We found many instances in which rather small new investments could increase productivity dramatically, and still they were not being made. The social costs resulting from the lack of access to capital faced by many small entrepreneurs are undoubtedly very high.

While talented and committed (particularly black) entrepreneurs may eventually arise, it might be socially desirable to help that occurrence by programmes that aim at detecting entrepreneurial talent and fostering its development. The United Nations program EMPRETEC tries to do that plus provide motivation for the budding entrepreneurs to not give up in their efforts. Another service of potential usefulness is help in preparing business plans and investment projects. Both EMPRETEC and SEDCO provide training and other support in this area, and it would be worthwhile to carefully examine the quality and reach of their operations to determine the possible need for additional assistance to these programmes.

Infrastructure

Transportation, electricity and water and sewerage are services in which the private sector generally cannot alone produce effective markets. In many cases, infrastructure projects generate large externalities not captured by private investors, their maturation takes too long for the relatively undeveloped state of local capital markets, and the size of efficient investments generally exceeds the capacity of the domestic private sector to fund them.

The experience of other developing countries, for example with electricity, shows that when the government fails to provide an adequate environment for large (economic) private or public energy plants to operate, the private sector will spontaneously create an 'informal sector' of electricity generation using very small plants at very high economic costs. That is, if the government does not succeed in establishing the basis for efficient production, the private sector alone will not achieve social efficiency. The Zimbabwean experience with the supply of electrical power and water during recent droughts also illustrates this point.

As the economy becomes increasingly more open, transportation will also become more critical, particularly for a landlocked country such as Zimbabwe, and the need for a government policy framework to encourage an efficient transportation system will increase. Another area of concern for industrial entrepreneurs is the telephone service, which is now unreliable and very expensive, particularly for international calls which will be increasingly important with the move towards a more open economy and more export-oriented manufacturing sector.

12.2 Human resources policy

High-level technical skills

Lack of high-level technical skills was apparent in our survey of Zimbabwean firms. Moreover, the number of engineers in Zimbabwe's manufacturing is very low compared to other countries at similar levels of development. As shown in Chapter 10, even Kenya, with lower industrial development, has a higher proportion of engineers in the manufacturing workforce. Until recently, Zimbabwe's industrial firms had little or no need to innovate processes or to improve quality and product design, as is routinely required in order to sell in highly competitive markets. They could do quite well as long as they were able to repair their machinery and produce goods of average quality. Thus the need for highly trained technical personnel was not perceived. If Zimbabwe is expected to accelerate its growth via an increasingly export-oriented manufacturing sector, it will require more university-trained engineers working in its plants.

Nevertheless, the 'purist' may still ask, why intervene? With freedom of entry and exit, the market for engineering services will be regulated by supply and demand. If firms do not hire engineers, or higher-level technicians, it is because they do not need them or their price is too high. If there is truth to this line of reasoning, it becomes important to determine why engineers may apparently not be needed, and also whether there is indeed a significant discrepancy between their social and market wages.

In small firms, explicit demand for engineers generally does not exist; does that mean that their services are not needed? No, it rather means that there is no appropriate organizational structure for an engineer to be included in a small firm, and that his services are indivisible, i.e. it is not feasible to hire one-quarter of an engineer. While it might be argued that consulting services and temporary workers exist for this reason, the argument may be irrelevant in developing countries such as Zimbabwe with weak and incomplete markets, in particular, for highly-skilled technical personnel services. Another possible cause for the lack of engineers demand is the low level of formal education of most managers. It is likely that in small, family-controlled plants, management feels uncomfortable hiring highly-trained personnel. In such a case, the private costs of hiring this kind of personnel are perceived to be much higher than the expected benefits. Of course, social costs are lower and benefits higher than their private counterparts.

Taken at face value, the high private cost argument seems to imply a situation of excess supply which would tend to be apparently confirmed by the information gathered by the survey team about recent university engineering graduates that could not find a job in industry. For those cases where hiring an engineer would be feasible, i.e. he/she could be satisfactorily integrated in the firm's existing organizational structure, the solution is of course to bring more in line social and market wages. Engineers may still command the remuneration they aspire to in view of their investment in training, but the government, an association of industrialists, or a combination of both, could provide a subsidy to hiring firms that will reduce their actual cost of employing engineers. It would work in a manner similar to the government's ZIMDEV (Zimbabwe Manpower Development Fund) programme for operative workers that subsidizes the first two years of in-plant training.

Such a measure must of course be predicated on the assumption that the argument offered by some industrialists that they do not hire engineers graduated from the local university because they are an inappropriate good due to their total lack of experience, is not the key, or most important, explanation. The level of the subsidy will have to be determined by trial and error under real conditions prevailing, in particular, in medium size enterprises which are the most likely initial candidates for such a programme.

Additionally, if the claims about the negative effect of lack of experience have validity, a programme of internships could be started during the last years of engineering training. In trying to address this need, the local association of engineers has been promoting a 'mentorship' programme whereby senior, retired engineers, would serve as tutors for young graduates to ease their incorporation to functioning manufacturing plants.

Lower-level technical skills

Manufacturing in Zimbabwe also employs few technicians and well-trained workers. Managers often prefer to employ untrained rather than trained and otherwise educated workers (see Chapter 8, section 6). A number of these managers base their decision on the belief that the education system does not produce useful graduates. This conviction has, for example, induced the **textiles and clothing** producers association to start their own training facility.

It seems important to strengthen the relationship between existing public training programmes and private sector users. The scope for intervention in this field is quite large, and there are a number of developing country experiences from which to draw. For possible immediate consideration would be to allow technical training institutions that are funded with the education levy from factory payrolls, to receive at least part of it directly, and to let industrialists have a say about the way in which these funds are spent. Under the current centralized budgetary system, if a public training institution such as Harare Polytechnic co-operates with the private sector and uses its training to generate revenue, it cannot keep any part of it. Thus, it does not have direct incentives to cooperate with the private sector. Such kind of arrangements need to be reconsidered. The goal would be to give public technical training institutions, such as Harare Polytechnic, whose equipment is very old and some of it not operable, a stable source of funding to allow them to modernize their equipment, to give potential users the chance to provide an early input into the educational product they are expected to buy, and to provide the training institutions the incentives to engage cooperatively with the private sector.

12.3 Trade policy

Liberalization

Although trade policy theory shows that domestic distortions are best addressed at their origin and not via protection, institutional restrictions and high transaction costs prevailing in developing countries such as Zimbabwe, frequently make it more costly to do so.

Additionally, opening up to international trade should not be indiscriminate, and a shock treatment is not always the best approach specially when compensating mechanisms to help those most affected are not in place.[38] Unrestricted imports of clothing and other textile products, for example, have had deleterious effects on many domestic factories unable to compete with internationally priced goods.

Moreover, technological learning by doing requires the opportunity to produce, and the attainment of long-term comparative advantage may need to be nurtured in some sectors. Otherwise, although potentially available, it may never get a chance to show up (see Westphal, 1982).

Regional issues and cooperation are quite important for a landlocked country such as Zimbabwe, in particular its relationships with South Africa its most important trade partner. South Africa is the source of more than 30% of its imports but the destination of only 12% of its exports. Although a trade agreement has been reached recently between the two countries, some of Zimbabwe industries most affected by the liberalization are claiming dumping.

Manufacturing exports

In addition to opening up to imports, trade policy will have to include the active promotion of manufactured exports. This is done not only in industrializing countries, but also in most industrialized countries. Besides maintaining a real exchange rate appropriate for industrial exports, programmes may include various types of incentives such as tax draw-backs, export credits, duty-free imports for exports, etc.

The expansion of exports will permit many plants to improve capacity utilization rates and eventually attain economies of scale, and also require that they meet more stringent cost and quality standards. This would demand technology and skills upgrading in a number of cases and result in beneficial externalities for the rest of the firms in an industry.[39]

Administration

To be effective, trade policy should be considered jointly with industrial and technology policy as has been done recently in successfully industrializing and industrialized countries as for example, Japan, Brazil, Korea and Taiwan.

Efficient and transparent administration of trade policy will demand the services of a highly qualified and fairly independent government bureaucracy.

12.4 Technological policy

Technology transfer

The absorption of foreign technology requires complementary local technological development. Reliance on direct foreign investment does not guarantee successful incorporation of technology if local skills and institutions are not available for its assimilation, adaptation, and eventual improvement.[40]

Little use is made by Zimbabwe's industrial establishments of foreign disembodied technology. Its application, via licences, technical assistance agreements, and expatriates, should be promoted via tax credits, subsidies for trips abroad to fairs and exhibitions, the easing of restrictions, etc.

Technical operating capabilities

Because a substantial proportion of the production equipment is old, maintenance and repair, as well as capital investment for eventual replacement, are required. The investment could be facilitated by means of accelerated depreciation rules and tax credits. Although Zimbabwe's entrepreneurs have had good experience in keeping their equipment running under blockade conditions, new competitive conditions require improved technical capabilities including the adoption of preventive maintenance routines.

To improve the **maintenance and repair** of machinery, as well as the upkeep of plant facilities, the deployment of more and better trained technical personnel is needed, in particular, in small and medium size firms.

Given the increasing importance of quality requirements when industry shifts more towards export markets, **quality control** methods will need to be improved. Concern for quality control requires that top management be imbued of its importance, and that more and better training of personnel at all levels be imparted. Greater awareness about consumer reactions and use of this feedback is necessary. It could be stimulated by publicity, training, etc. The Standards Association of Zimbabwe (SAZ) will need to expand its activities of norms formulation, quality testing, and training and technical assistance for the implementation of ISO 9000 and other international quality standards.

Productivity improvement and marketing techniques are also important in the context of an export-oriented industrial drive, and may require appropriate international assistance.[41] The role that **industrial engineering** techniques could play in productivity improvement, across industries, needs to be stressed and publicized. More industrial engineers (of the US variety) should be trained. Courses on industrial engineering techniques such as work-study methods, layout planning, work sampling, optimization of scheduling and inventory control, cost control, etc., should be imparted for management and other personnel. The use of computers for scheduling and the integrated planning of production and inventory control should be promoted.

The importance of **industrial safety** needs to be highlighted. Safety training should also be offered, as well as incentives for the use of protective equipment in hazardous tasks. Insurance underwriters should be stimulated to grant premium discounts in their policies for factories complying with higher safety standards. Such standards should be developed with the collaboration of workers' and industry representatives. Adoption of a policy for vigorous equipment replacement, will, as a byproduct, also result in a machinery park with better means for worker protection.

Greater awareness about the significance of **environmental control** in modern manufacturing industry needs to be attained in Zimbabwe. Publicity and training should be used to that effect. The government should cooperate with the private sector in solving difficult industrial effluent disposal problems by facilitating access to technical assistance. It should also step up its controls and punish those firms not taking measures to solve routine effluent disposal problems. A policy of accelerated equipment replacement will also help in environmental control, because antipollution devices are generally incorporated in more modern equipment.

Industrial research and experimental development

The structuring of formal technological activities at the firm level starts with the allocation of a minimum amount of resources and a place in the organization. The existence of technical documentation and of certain key technical offices are good indicators of the significance assigned in a firm to R&D and related technical activities.

Incentives could be provided for firms undertaking applied R&D in an organized and systematic fashion, including support for the hiring of highly skilled personnel as well as the contracting of outside R&D work. Small and medium size firms, and those without foreign or state ownership, are generally unaware of the research services they may need, and get, and, for similar reasons, also hesitate in hiring technical personnel, or contracting outside research (see World Bank, 1993c).

There are no rigid rules permitting a clear distinction between basic research, admittedly a public good, and applied research, which may have characteristics of both private and public good (Nelson, 1992). Largely because of a desire to support industrial R&D in certain leading new sectors, there has been a relaxation as to what is acceptable public policy in this area in a number of countries. Thus fairly widespread acceptance now exists, even among industrialized countries, for the notion that R&D is to a large extent a public good and consequently deserves to be subsidized to bring its social and market costs and benefits into convergence.

Programmes in other countries have tried to address the problem of a lower than optimal private investment in R&D due to public good characteristics of research such as incomplete appropriability and uncertainty of the results. Successful programmes may include risk-sharing, joint public-private undertakings, and public sponsoring of some research projects, in addition to tax benefits for R&D expenditures similar to those granted for capital investment.[42]

Agreements reached in the Uruguay Round of the GATT and creation of the World Trade Organization (WTO) have established the principle that research subsidies are permitted. Up to 75% of the cost of basic research, and 50% of that of applied research, could be subsidized via government programmes. Still, the problem arises as to how best to do it, particularly in economies with a very limited scientific and technological basis, weak institutions, and a scant cadre of high-level research personnel.

In some countries governments have created technological research institutes but in a number of cases it has been argued that their research programmes do not take into account the needs of industry, giving rise to the concern of making their work relevant to industry. However, institutes such as KIST in Korea, and Instituto de Pesquisa Tecnologica (IPT) in San Pablo, Brazil, are well established and have a pretty good record. Support for the creation of the various institutes planned for the SIRDC should be forthcoming from local and international donors, and measures should be taken to assure, from the very beginning, that its work is relevant to the needs of Zimbabwe's manufacturing sector.

A word should be said about public R&D. The university needs to adopt a scheme similar to that used in good universities in industrialized countries, whereby teaching and research go hand in hand and professors are rewarded for their ability to combine both. Given the recent astounding growth of the student population, measures to preserve the quality of instruction while improving the level of the staff, and the equipment available in laboratories and other facilities, will be difficult to implement by a government financially constricted and embarked in an austerity programme. International aid, carefully coordinated and executed, could make an important contribution in this area.

The proportion of scientists and engineers employed in industry is too low, and the results of their research are too concentrated in biology and medicine to the detriment of engineering, technology, and related sciences. The use of public incentives to achieve a change in the allocation of science and technology resources in the direction of favouring the technology-related scientific fields should be actively considered.

Appendix A: Questionnaire

BACKGROUND

		Tekere	Mumvuma	Thoumi	Teitel
1. Date:	2. Interviewer(s):	❑	❑	❑	❑

3. Name of Person Being Interviewed: ______________________

4. Position: ______________________

5. Name of the firm: ______________________ 6. Industry code: ❑ 31 ❑ 32 ❑ 33 ❑ 38

7. Address: ______________________

Tel: ______________________ Fax: ______________________

8.

Principal Activities	description	% of total revenue
(1) Production		
(2) Repair & Spares		
(3) Imports & Sales		
(4) Sale of local 3rd party goods		
(5) Other		

9.

Principal Products	% of total sales	% exported
(1)		
(2)		
(3)		
(4)		
(5)		

10. Type of Production

a) To order ❑
b) For stock and distribution ❑
c) Both ❑

11. Market Orientation

a) 100% Domestic ❑
b) < 10% Exports ❑
c) 10%–25% Exports ❑
d) 26%–50% Exports ❑
e) > 50% Exports ❑

Comments/notes: ______________________

____

____

12. What is the sequence of manufacturing operations in your plant? For the main product, please provide the names of all separate manufacturing operations and a brief description of each. (For example, a typical sequence in some metalworking industries is: casting, forging, heat and surface treatment, testing, specialized machining and final machinery assembly.)

____

____

____

____

____

____

____

____

____

13. Indicate below the category best representing your manufacturing operation.

(a) Processing raw materials only to semi-processed or commodity level ❑ 1

(b) Production of components for assembly by a third party ❑ 2

(c) Assembly of final product (only) ❑ 3

(d) Full sequence manufacturing operations up to final product ❑ 4

(e) Other (specify) ______________________________ ❑ 5

(f) Combination ❑ ❑ ❑ ❑ 6

Comments/notes: ______________________________

____

14. What is the organizational structure of your firm/plant? Draw a chart indicating the hierarchical levels and the principal departments and functions: production, sales, finance, personnel or industrial relations, controllership, engineering, quality control, research, etc.

15. Professional Management

a) Owner is General Manager ❑
b) There is professional management ❑
c) Other (explain) ❑

____

____

16. Type of Organizational/Structure

a) Manager directs workers ❑
b) Manager directs foremen who direct workers ❑
c) One tier management with separate functions ❑
d) Several management tiers with functional division of responsibilities ❑
e) Other (explain) ❑

____

____

17. Existence of Formal Individually Assigned Management Functions

		Yes	No
(a)	General Manager	❑	❑
(b)	Sales Marketing	❑	❑
(c)	Finance	❑	❑
(d)	Production	❑	❑
(e)	Administration	❑	❑
(f)	Personnel	❑	❑
(g)	Purchasing	❑	❑
(h)	Maintenance	❑	❑
(i)	Quality Control	❑	❑
(j)	Technical	❑	❑
(k)	Plant	❑	❑
(l)	R&D	❑	❑
(m)	Engineering	❑	❑
(n)	Others (specify)	❑	❑

____

____

I. TECHNOLOGICAL HISTORY OF THE FIRM OR PLANT

INITIAL TECHNOLOGY

1. How was the firm/plant started?

	Preinvestment or Project Study Yes	No
(a) Started as very small operation; single or few machines	❑	❑
(b) Started investing in full-scale plant	❑	❑
(c) Current owner acquired going concern	❑	❑
(d) Foreign investment or joint venture	❑	❑

If No, go to question 3 ⇨

2. For yes answer above, who carried out the technical feasibility study for the factory?

(a) Entrepreneur ❑ 1
(b) In-house staff ❑ 2
(c) Local consultants ❑ 3
(d) Expatriate consultants ❑ 4
(e) Foreign partner or headquarters ❑ 5
(f) Other (specify) ❑ 6

____

____

(g) Combinations ❑ ❑ ❑ ❑ 7

3. Who selected the equipment?

(a) Entrepreneur ❑ 1
(b) In-house staff ❑ 2
(c) Local consultants ❑ 3
(d) Expatriate consultants ❑ 4
(e) Foreign partner or headquarters ❑ 5
(f) Other (specify) ❑ 6

____

____

(g) Combinations ❑ ❑ ❑ ❑ 7

4. What role did the firm play in specifying the equipment purchased?

(a) None ❑ 1
(b) It was fully specified by the firm ❑ 2
(c) Interfaced with supplier in producing full specification for equipment ❑ 3
(d) Received specifications from headquarters or foreign partner ❑ 4
(e) Other (specify) ❑ 5

____

____

(f) Combinations ❑ ❑ ❑ ❑ 6

5. How did the firm locate its equipment?

	Yes	No
(a) Trip abroad	❑ 1	❑
(b) Catalogue	❑ 2	❑
(c) Saw similar machines in fair/exhibition	❑ 3	❑
(d) Other (specify)	❑ 4	❑

____

____

6. Provide information about the name, type, country of origin and average age of the three principal items of equipment in the factory:

Name of Equipment	Function	Country of origin	Age	Bought New or Used
(a)				
(b)				
(c)				

7. What is the replacement value of your current equipment? $____________

8. Amount of plant insurance? $____________
(Breakdown into building and equipment)

9. What is the power installed (H.P.)? ____________

10. If you were to buy new equipment for your factory, to produce the same product(s), would you buy the same or different machinery? Explain:

(a) Would buy practically the same	❑ 1
(b) Would buy equipment with same production capacity but new technology/process	❑ 2
(c) Would buy higher capacity, new technology	❑ 3
(d) Would buy higher capacity, same technology	❑ 4
(e) Other (specify)	❑ 5

____

____

11. Reasons for answer in question 10.

	Yes	No
(a) To lower costs by increasing productivity	❑	❑
(b) To change product line	❑	❑
(c) To improve quality	❑	❑
(d) Other (specify)	❑	❑

____

____

(e) Combinations ❑ ❑ ❑

12. Are you using foreign licenses and/or technical assistance?

Yes ❑ 1

No ❑ 2 *If No, go to question 15* ⇨

13. What were the contractual terms and forms of payment for the technology transfer?

	Yes	No
(a) Lump sum fee	❑	❑
(b) Royalty (% of sales or profits)	❑	❑
(c) Technical assistance agreement with a fee for service	❑	❑
(d) Other (specify)	❑	❑

Comment: ______________________________

____

14. If you had a contract, did it contain any of the following restrictions:

	Yes	No
(a) Tie-in clauses for imports of materials and components	❑	❑
(b) Export restrictions	❑	❑
(c) Provisions for cross-transfer of know-how	❑	❑
(d) Other restrictions or limitations (specify)	❑	❑

____

____

15. Did you receive any help for commissioning the plant or installing the equipment?

Yes ❑

No ❑ *If No, go to question 17* ➪

16. Did it involve any of the following?

	Yes	No
(a) Technical personnel from supplier come to install equipment	❑	❑
(b) Your personnel were trained to operate the equipment	❑	❑
(c) Your personnel went abroad to train in the operation of the equipment	❑	❑
(d) Other support (specify)	❑	❑

____

____

17. What kind of <u>technical</u> problems did you face when you started production? List the most important problems and describe how they were solved.

Technical problem	**Describe steps taken to solve**
(a) Failure to reach full capacity allowed by equipment design. How long did it take?	
(b) Quality Control: 1. Meeting standard product specifications	
2. High proportion of rejects	

continued...

Technical problem	Describe steps taken to solve
(c) Meeting process specifications:	
(d) Equipment Breakdowns:	
(e) Other (describe)	

18. Overall, of the categories given below, which best represents your start-up trouble-shooting experience?

    (a) Minor problems, easily overcome by training and technical support ☐ 1

    (b) Major problems, required extensive experimentation and/or technical support ☐ 2

    (c) Trouble-shooting failed, firm gave up, reduced range of activities, and/or bought other equipment. ☐ 3

    (d) Other (explain) ☐ 4

    ____

    ____

II. OPERATING CAPABILITIES

In this part of the survey we try to reflect the key technological characteristics of the firm and its main plant as to : equipment and buildings, organization, and production capabilities.

GENERAL

1. How does the plant look? (well organized, clean, safe, well lighted, etc.) Describe:

____

____

____

____

2. Type of production facility

(a) Premises built for, or well adapted to, industrial production; well kept, clean, lighted, etc. ❑ 1

(b) Industrial premises as above but poorly kept ❑ 2

(c) Mixed sites, yards, open plus roofed areas, heaps of old, unused materials, old equipment, mixed with current production and inventories. ❑ 3

(d) Minimal production facilities attached to retail shop, houses, offices, street, etc. ❑ 4

3. Is the equipment mostly new or mostly old? Describe:

____

____

____

____

4. How many shifts does the plant work: ❑ 1 ❑ 2 ❑ 3

5. Is that the industry standard? ❑ yes ❑ no

Explain:

6. What is the rate of capacity utilization, per shift? _______ %

7. Are there idle machines or any other unutilized equipment?

Comment: _______________

____

____

____

____

The remainder of this section includes consideration of the following:

A. Capacity to troubleshoot production processes

B. Capacity to control quality

C. Capacity to maintain and repair the production equipment

D. Capacity to improve production processes

E. Industrial safety and pollution control

A. TROUBLESHOOTING

1. Which of the following technical problems did the plant experience during the last 2 years:

(Do not include problems due to general economic conditions or regulations, such as lack of foreign exchange, or imported spare parts, etc.)

	Yes	No
(a) Equipment breakdowns	❑	❑
(b) Serious quality control problems	❑	❑
(c) Bottlenecks (imbalances) in the production process	❑	❑
(d) Safety problems (accidents)	❑	❑
(e) Pollution control difficulties	❑	❑
(f) Other (specify)	❑	❑

____

____

2. If the answer to any of the above is yes, please explain how the problem was solved in each case:

Problem: ______________________________

	Yes	No
(a) By in-house personnel	❑	❑
(b) With outside local help	❑	❑
(c) With outside foreign help	❑	❑

In each case provide details below:

____

____

____

____

____

Problem: ______________________________

	Yes	No
(a) By in-house personnel	❑	❑
(b) With outside local help	❑	❑
(c) With outside foreign help	❑	❑

In each case provide details below:

____

____

____

____

____

continued...

Problem: ______________________________

	Yes	No
(a) By in-house personnel	❑	❑
(b) With outside local help	❑	❑
(c) With outside foreign help	❑	❑

In each case provide details below:

____

____

____

____

____

B. QUALITY

1. Indicate the quality control system used for:

	None	Visual, practical, inspection	Laboratory tests, use of standards, no statistics	Same as before, plus statistical methods
(a) Final product				
(b) Process(es)				
(c) Raw materials and components purchases				
(d) Materials and components in process				

2. Do you use any national or international standards and norms for the final product or raw material inputs?

	Yes	No
(a) National norms	❑	❑
(b) International norms	❑	❑

3. If the answer to 2a or 2b is yes, describe and state when they were adopted:

____

____

____

4. If the answer to 2a or 2b is yes, is their use:

(a) Compulsory ❑ 1

(b) Voluntary ❑ 2

5. Is there any established routine for testing during the production process by taking samples for inspection?

Yes ❑ 1

No ❑ 2

6. How good is your quality compared to the best local competitor and/or imports? Rank your quality by picking one of the following:

	Local	Imports
(a) Much higher	❑	❑
(b) Higher	❑	❑
(c) Equal	❑	❑
(d) Lower	❑	❑
(e) Much lower	❑	❑
(f) Do not know	❑	❑

7. What were the internal reject rates in the manufacture of your main product? (Reject includes reworking, selling at lower price, selling under second brand name, etc.)

	%
(a) Last year	
(b) 3 years ago	
(c) 5 years ago	
(d) Not applicable (production to order, or other reason)	Explain:

If there is an improving trend, explain how and when it happened.

____

____

____

8. How do you keep track of client (customer) complaints? Describe:

(a) No methods to keep track ❑ 1
(b) By direct contact with customers ❑ 2
(c) Through sales force ❑ 3
(d) Other (explain) ❑ 4

____

____

9. Do you have a system of 'statistical quality control'?

Yes ❑ 1
No ❑ 2

C. MAINTENANCE AND REPAIR

1. Do you have an in-house maintenance and repair shop, and/or qualified staff to do the job?

   Yes ☐

   No ☐ *If No, go to question 5* ⇨

2. Is there an established routine for preventive maintenance of equipment and machinery?

   Yes ☐

   No ☐

3. When do you stop to check machines?

   (a) Only when something breaks down ☐ 1

   (b) When wear is perceived (e.g., noises or vibrations are heard) ☐ 2

   (c) At established times for check-up, whether there are problems or not ☐ 3

   (d) At established times for scheduled replacement of parts ☐ 4

   (e) Other (specify) ☐ 5

   ____

   ____

4. Frequency of servicing (oiling, etc.): ____

   ____

   Frequency of parts replacement: ____

   ____

5. What outside help is used?

   Kind: ____

   ____

6. Are equipment breakdowns frequent?

   Yes ☐ 1

   No ☐ 2

7. How often: ____

   ____

8. Who is in charge of repair and maintenance?

   ____

   ____

9. Who fixes the machines when they break down?

   (a) Totally dependent on outside help ☐ 1

   (b) Self servicing, oiling, replacement of belts or similarly accessible items ☐ 2

   (c) Same as (b) plus capacity to strip most machines to replace standard parts or getting spares made ☐ 3

   (d) Complete service of repair and maintenance in house ☐ 4

   (e) Other (explain) ☐ 5

   ____

   ____

D. INDUSTRIAL ENGINEERING

1. Is there a continuous, formally organized effort to improve plant productivity without the introduction of new equipment?

Yes	❑ 1	
No	❑ 2	Go to question 4 ⇨

2. How many of your personnel are <u>permanently</u> assigned to this task?

1–5	❑ 1
6–10	❑ 2
> 10	❑ 3
None	❑ 4

3. What is the nature of their qualifications? How many are:

<u>Engineers</u> *(university graduates)*	____________
<u>Technicians</u> *(diploma/certificate)*	____________
<u>Practically trained technicians</u> *(foremen/workers)*	____________

4. Do you carry out time and motion studies?

Yes	❑ 1	
No	❑ 2	Go to question 8 ⇨

5. What method(s) is(are) used to set production standards? How frequently are standards set? Describe.

____

____

6. How many personnel are assigned to this task? ____________

7. What are their qualifications?

____

____

8. How do you compare your productivity (output per employee) to that of other firms in the same industry? (Output could be measured in money terms or in physical units.)

	Higher (%)	Same	Lower (%)	Don't Know
Small local	❑ __	❑	❑ __	❑
Large local	❑ __	❑	❑ __	❑
Foreign	❑ __	❑	❑ __	❑

9. Please indicate whether advance scheduling of production is carried out in the plant/firm:

Yes	No
❑ 1	❑ 2

10. If the answer to 9 is yes, please indicate the type of scheduling:

(a) Day-to-day	❑ 1
(b) System based on delivery dates established for orders	❑ 2
(c) Forward planning at least for one week to meet production or stock targets	❑ 3
(d) Longer term planning for internal or export markets, at least one month ahead, probably much more	❑ 4

11. Please indicate whether the following functions are carried out in the plant/firm:

Formal inventory control of:	Yes	No
a) raw materials	❑	❑
b) work in progress	❑	❑
c) final product	❑	❑

12. Are computers used in carrying out any of the above?

Yes	❑ 1
No	❑ 2

Describe (hardware/software): ____________________

____

13. Is the plant layout continuously being assessed and improved?

Yes	❑ 1
No	❑ 2

14. Is the plant layout reexamined only on occasion of plant expansion or replacement of equipment?

Yes	❑ 1
No	❑ 2

15. When was the layout of machines in the plant last changed? Why, and who did it?

____

____

____

____

E. INDUSTRIAL SAFETY AND POLLUTION CONTROL

1. Is the responsibility for industrial safety assigned to any particular office?
   - Yes ☐
   - No ☐

Describe. ____

____

2. Is the responsibility for environmental (pollution) control assigned to any particular office?
   - Yes ☐
   - No ☐

Describe. ____

____

3. Are there written rules for safety?
   - Yes ☐
   - No ☐

4. Are there written rules for environmental control?
   - Yes ☐
   - No ☐

5. Are there periodic drills for fire evacuation and other safety procedures?
   - Yes ☐
   - No ☐

6. How do you rate your safety record and procedures compared to other firms in your industry?
   - (a) Higher ☐
   - (c) About the same ☐
   - (d) Lower ☐

7. How do you rate your pollution control record and procedures compared to other firms in your industry?
   - (a) Higher ☐
   - (c) About the same ☐
   - (d) Lower ☐

III. TECHNOLOGICAL FUNCTIONS AND ORGANIZATION

A. GENERAL

1. Is the technology (i.e., technical information to operate) largely codified in manuals, instructions, blueprints, specifications, etc., or is it tacit and diffused by word of mouth methods via supervisors and other personnel? Describe the basic technical documentation if any.

____

____

____

____

2. Mark the offices that exist in the firm/plant and indicate their number of employees on the respective line.

		Number
❑	Product Engineering	______
❑	Manufacturing or Process Engineering	______
❑	Quality Control Laboratory	______
❑	Industrial Engineering	______
❑	Research and Development	______
❑	Pilot or Experimental Plant	______

3. Other technical offices and number of employees:

____

____

____

____

4. Formalization of technical knowledge in use and technical offices available.

(a)	No documentation or files, no technical offices	❑ 1
(b)	Very little documentation, isolated drawings, charts or sketches; and/or up to two technical offices	❑ 2
(c)	Good technical documentation and/or three or more technical offices	❑ 3

B. PRODUCT DESIGN AND CHANGES

1. Origin of the main product's design:
   - (a) Foreign specification ❑
   - (b) Self-generated, domestic specification ❑
   - (c) Copy of similar imported product ❑
   - (d) Copy of similar domestically produced product ❑
   - (e) Other (specify) ❑

   ____

   ____

2. Changes or modifications introduced:
   - (a) None ❑ *Go to question 4* ⇨
   - (b) Some, but not very significant ❑
   - (c) Substantial changes ❑

3. If changes were made to the product design, indicate which of the following apply:
   - (a) Product modification done to suit local tastes ❑ 1
   - (b) Product modification done to suit local climate ❑ 2
   - (c) Product modification due to use of local materials ❑ 3
   - (d) Product modification done to simplify it ❑ 4
   - (e) Product modification due to use of available equipment ❑ 5
   - (f) Product modification to update design ❑ 6
   - (g) Modification done to differentiate product ❑ 7
   - (h) Product modification due to meet market needs or to incorporate new functions ❑ 8
   - (i) Product modification due to other reasons (specify) ❑ 9

   ____

   ____

   - (j) Combinations ❑ ❑ ❑ ❑ 10

4. Introduction of new products

   Were any new products introduced in:

	Yes	No
(a) The last year	❑	❑
(b) The last 5 years	❑	❑

If either (a) or (b) is answered yes, go to question 5.

If both (a) or (b) are answered no, go to section C.

5. Please provide the following information:

	Yes	No
(a) Were they developed in-house?	❑	❑
(b) Was any outside help provided for the design?	❑	❑
(c) Was any outside help provided for the testing of prototypes or samples?	❑	❑

(d) Describe below any outside services that participated in the design and development of the new product(s):

____

____

____

____

____

____

____

C. MODIFICATIONS AND IMPROVEMENTS IN THE PRODUCTION PROCESS

1. Did you carry out any of the following technical changes:

	Yes	No
(a) Down-sizing of the production process	❑	❑
(b) Adaptation to local raw materials	❑	❑
(c) Energy saving	❑	❑
(d) Capacity stretching	❑	❑
(e) Manufacture of new tools/dies/fixtures	❑	❑
(f) Development of new processes	❑	❑
(g) Other technical changes (specify)	❑	❑

____

____

2. For each of the technical changes noted affirmatively above, please provide the following information:

Technical change ______________________ Date __________

i) Reasons for the change:

____

____

ii) Expected benefits:

____

____

iii) Cost:

____

____

iv) Personnel that carried it out:

____

____

v) How long did it take until fully implemented?

____

____

vi) Were the expected benefits attained? Explain in each case the main discrepancies with your expectations.

____

____

vii) Were additional unexpected benefits realized and/or costs incurred? Please explain in each case.

____

____

Technical change ______________________ Date __________

i) Reasons for the change:

____

____

ii) Expected benefits:

____

____

iii) Cost:

____

____

iv) Personnel that carried it out:

____

____

v) How long did it take until fully implemented?

____

____

vi) Were the expected benefits attained? Explain in each case the main discrepancies with your expectations.

____

____

vii) Were additional unexpected benefits realized and/or costs incurred? Please explain in each case.

____

____

Technical change _______________ Date _______________

i) Reasons for the change:

____

ii) Expected benefits:

____

____

iii) Cost:

____

____

iv) Personnel that carried it out:

____

____

v) How long did it take until fully implemented?

____

____

vi) Were the expected benefits attained? Explain in each case the main discrepancies with your expectations.

____

____

vii) Were additional unexpected benefits realized and/or costs incurred? Please explain in each case.

____

____

D. SUPPORT SERVICES

1. Did you receive support services for any of the following technical activities?

	Yes	No
(a) Quality control (testing or certification)	❑	❑
(b) Repair and maintenance	❑	❑
(c) Technical assistance	❑	❑
(d) Training of personnel	❑	❑
(e) Instruments calibration	❑	❑
(f) Other (specify)	❑	❑

____

____

2. For each yes answer above provide below the following information:

Service: ______________________________

a) Nature of service:

____

____

b) Provider of service:

____

____

c) How frequent: d) Payment

______________ ______________

______________ ______________

e) Satisfied? Yes ❑ 1 No ❑ 2

If No, please explain why:

____

____

Service: ______________________________

a) Nature of service:

____

____

b) Provider of service:

____

____

c) How frequent: d) Payment

______________ ______________

______________ ______________

e) Satisfied? Yes ❑ 1 No ❑ 2

If No, please explain why:

____

____

continued...

Service: ______________________________

a) Nature of service:

____

____

b) Provider of service:

____

____

c) How frequent:

____

____

d) Payment

____

____

e) Satisfied? Yes ❑ 1 No ❑ 2

If No, please explain why:

____

____

3. What technological support services would be beneficial for your firm if readily available and efficiently delivered? Assign priority from 1 to 5 with 1 being the highest.

	Yes/Rank	No
(a) Testing for quality control	❑	❑
(b) Personnel training	❑	❑
(c) Repair and maintenance	❑	❑
(d) Calibration of control instruments	❑	❑
(e) Technical assistance for troubleshooting	❑	❑
(f) Productivity improvement techniques (e.g., time and motion studies, layout changes)	❑	❑
(g) Energy saving	❑	❑
(h) Industrial safety	❑	❑
(i) Labor relations	❑	❑
(j) Pollution control	❑	❑
(k) Tools, dies or fixtures manufacture	❑	❑
(l) Product redesign	❑	❑
(m) Process improvement	❑	❑
(n) Pilot plant experimentation	❑	❑
(o) Other (specify)	❑	❑

____

____

E. TRAINING

1. Is on-the-job training required to operate the plant's equipment?

	Yes	No
(a) For formally trained workers	❑ 1	❑ 2

If the answer is yes, what is the duration of the training required to achieve satisfactory performance?

____

	Yes	No
(b) For workers with similar job/employment experience	❑ 1	❑ 2

If the answer is yes, what is the duration of the training required to achieve satisfactory performance?

____

	Yes	No
(c) For other workers	❑ 1	❑ 2

If the answer is yes, what is the duration of the training required to achieve satisfactory performance?

____

2. Is training for professional personnel undertaken by the firm?

Yes	❑ 1
No	❑ 2

If the answer is yes, provide the following information:

Reason for training need: ______________________________

Nature of training: ______________________________

Place of training (i.e., locally, abroad, etc.): ______________________________

3. Is training provided for technicians and other middle-level technical personnel?

Yes	❑ 1
No	❑ 2

If the answer is yes, provide the following information:

Reason for training need: ______________________________

Nature of training: ______________________________

Place of training (i.e., locally, abroad, etc.): ______________________________

IV. PLANS FOR TECHNOLOGICAL DEVELOPMENT

1. General

(a) Firm plans to stay as it is (no changes in capacity or technology ❑ 1

(b) Firm has doubts about the future, and is considering withdrawing from industrial production ❑ 2

(c) Firm would stay as it is (capacity-wise), but plans to upgrade its technology ❑ 3

(d) Firm is planning to expand and to invest in technological upgrading ❑ 4

If answer is 1 or 2, go to Skills Roster ⇨

2. Does the owner, or general manager, envisage the need for technology changes in the plant/firm?

Yes ❑ 1

No ❑ 2

If the answer is yes, in what areas:

____

____

3. Do the plans involve the following:

	Yes	No
(a) Acquisition of new equipment	❑	❑
(b) Hiring of technical personnel	❑	❑
(c) Contracting for specialized services	❑	❑
(d) Getting a manufacturing license	❑	❑
(e) Signing a technical assistance agreement	❑	❑
(f) Other (specify)	❑	❑

____

____

4. What are the specific motivations for such changes as are envisaged:

____

____

____

____

5. Will local technical services be hired?

Yes ❑ 1

No ❑ 2

If the answer is yes, which ones:

____

____

6. How will the proposed changes affect the operation of the plant?

	Yes	No
(a) New products	❑	❑
(b) Better quality of existing products	❑	❑
(c) New processes	❑	❑
(d) Higher productivity	❑	❑
(e) Cost reduction	❑	❑
(f) Subcontracting	❑	❑
(g) Other (specify)	❑	❑

____

7. Will training of existing personnel be necessary to carry out the planned technological changes?

Yes	❑ 1
No	❑ 2

If the answer is yes, provide the following information:

Type of training: ______________________________

Duration: ______________________________

Where imparted: ______________________________

8. How long a learning process is envisaged to reach design capacity with the new technology:

____

9. Will there be external assistance to debug and troubleshoot the new process(es)?

Yes	❑ 1
No	❑ 2

If the answer is yes, where will it come from:

10. Will organizational changes also be necessary?

Yes	❑ 1
No	❑ 2

If the answer is yes, of what kind:

New functions: ______________________________

New offices: ______________________________

New responsibilities within existing structure: ____________________

Major changes to organizational chart: ____________________

11. What will be the estimated cost of the planned technological change?

Machinery: ______

Buildings and structures: ______

Research and development personnel costs: ______

Other equipment: ______

Supplies: ______

Other (specify): ______

Total Cost: ______

(If breakdown is not available, supply the total cost figure only.)

12. Will the proposed technical changes affect the competitive position of the firm?

Yes	❑ 1
No	❑ 2

If the answer is yes:

13. How will the firm rate in the domestic market in terms of cost and quality once the planned changes are undertaken?

Cost-wise: ______

Quality-wise: ______

14. How will it compare with relevant international producers?

Cost-wise: ______

Quality-wise: ______

Notes: ______

SKILLS ROSTER Provide the number of personnel in each case (including managers) with the required breakdown by skill or training.

Professional Personnel (with university degrees)

Profession	Trained locally	Trained abroad	Expatriates	Total
Engineers				
Chemists				
Accountants				
M.B.A.'s				
Other (specify)				
Total				

Technical Personnel (with diploma or certificate)

Specialty	Diploma	Certificate	Expatriates	Total
Metalworking				
Woodworking				
Food technology				
Textiles				
Electrical				
Construction				
Electronic				
Data processing				
Other (specify)				
Total				

Workers

Specialty	
Machine-tool operator	
Tool and die-makers	
Mechanics	
Electricians	
Material Handling	
Janitorial and Cleaning	
Machine Operations	
Others (specify)	

Part-time Workers ________

Apprentices ________

Seasonal Workers ________

Casual Workers ________

Total number of employees:	

Appendix B: Firm Interviews by Sector: Food and Beverages (ISIC 31)

Under food and beverages the survey included firms belonging to ISIC 31 with the exception of tobacco manufacturers. A total of 13 food and beverages industry establishments were visited and its management interviewed. This represented 32% of the total sample of 41 establishments.

A. Background information

1. General characteristics

The firms visited vary significantly in terms of product mix, employment size, output per worker, ownership and ethnic origin; see Table A-1. While a majority were individually owned firms, the group also included a multinational

Table A-1 Food and Beverages. Basic characteristics of firms interviewed

Firm	*Product*	*Employment*	*Ownership/ ethnic*	*Location*	*Sales/worker (US$)*
11	Bread, confect.	126	White	Harare	5,000
211	Sausages	54	White	Harare	22,800
241	Sweets, choc. biscuits	365	White	Harare	18,800
361	Chips, coffee snacks	800	Group/White	Harare	15,000
551	Spices	49	White	Harare	7,000
591	Milk, cereals	150	MNC	Harare	76,200
1401	Bottling	352	White	Bulawayo	27,400
1661	Sweets, choc. biscuits	294	White	Bulawayo	8,200
1731	Packaging	21	White	Harare	8,200
71	Meat pack. proces. cann.	1131	Public co.	Harare	7,100
1221	Milk,cheese, etc.	839	State. Dairy Market. Board	Harare	55,200
1351	Bakery	4	Asian	Bulawayo	9,200
1951	Packaging	21	White	Harare	8,200

subsidiary, a member of a large local holding group, as well as a cooperative and a large para-statal firm.

The range of output per worker is indicative of the variation in the extent of value added locally and the degree of mechanization of the various installations. The range was from $5,000 to $76,200 with an unweighted mean of $22,740. Large firms tend to have higher output per worker values. Several of the firms interviewed are only engaged in limited processing, packaging and bottling operations, and their extent of value added is accordingly quite low; see Table A-2.

2. Market orientation

Exports as a proportion of total output are shown in Table A-3. Although this is an industry where Zimbabwe is expected to have comparative advantage based on natural resources and low wage labour, export performance is not very strong. Only three of the 13 firms export 10% or more of their output (and of these none more than 20%), while four firms have no exports at all. The main markets are neighbouring countries in Southern Africa.

3. Organization

Under this heading the following items were covered: whether there was professional management, the type of organizational structure, and which managerial functions existed and were individually assigned. Table A-4 deals with the first two issues.

As can be seen, only seven of the 13 firms have professional management at the top (i.e. other than the owner of the firm). Of the firms with professional management, five also have several tiers of functional management, while two only have one-tier management functions.

Table A-2 Food and Beverages. Extent of manufacturing value added

Firm	*Type of operation*
11	Full sequence manufacturing
211	Process., packag., manufacturing
241	Full sequence manufacturing
361	Full sequence manufacturing
551	Process., packag., manufacturing
591	Full sequence manufacturing
1401	Bottling
1661	Full sequence manufacturing
1731	Processing and packaging
71	Processing, packing and canning
1221	Milk processing
1351	Full sequence manufacturing
1951	Manufacturing and bottling

Table A-3 Food and Beverages. Proportion of output exported

Firm	*Exports (000's US$)*	*Turnover (000's US$)*	*% Exported*
591	1,429.0	11,429.0	12.5
551	21.0	342.0	6.1
361	128.6	12,000.0	1.1
241	137.2	6,860.0	20.0
211	0.0	1,230.0	0.0
11	0.0	634.3	0.0
1731	4.3	171.4	2.5
1661	255.0	2,571.0	10.0
1401	0.0	9,857.0	0.0
71	n.a	8,000.0	<10.0
1221	n.a	46,300.0	<10.0
1351	0.0	37.0	0.0
1951	554.5	11,090.0	5.0

A strong tendency was observed among white individual entrepreneurs to concentrate managerial decision-making around themselves.

Firms that have several tiers of functional management also tend to have a larger number of such functions individually assigned. For example, a subsidiary of a MNC in this industry has general manager, sales, finance, production, personnel, purchasing, quality control and plant engineering functions assigned to specific individuals. A small firm, with no professional in the top management job, but one-tier functional management, had a sales

Table A-4 Food and Beverages. Organizational structure

Firm	*Profess. manag.*		*Organizational structure*				
	Yes	*No*	*Direct*	*Super.*	*1 Tier*	*Tiers*	*Other*
591	X					X	
551		X			X		
361	X					X	
241		X			X		
211	X				X		
11		X			X		
1731		X			X		
1661		X			X		
1401	X					X	
71	X					X	
1221	X					X	
1351		X	X				
1951	X				X		

manager, but the managing director also intervened directly in sales; had a production manager, but no finance manager because the managing director himself carries out these functions. The personnel function is absorbed by the plant manager and while there is somebody assigned to purchasing, the managing director also intervenes. In a smaller firm, without professional top management, there is an accountant also in charge of finance and the payroll, a sales manager also in charge, partially, of purchasing, and a foreman in charge of production. The owner-managing director also participates in purchases.

Thus size of firm is clearly limiting the extent of functional division of labour internal to the firm.

B. Entrepreneurial history and acquisition of technology

1. Origin of the firm and initial technology

Information was not always available about the starting up of the firm. In such cases, reference was made to the most recent change in ownership about which our respondent had information.

Firm 11. Original entrepreneur, father, is from Italy. Went to Somalia, then an Italian protectorate, and worked there until 1956. Then moved to Zambia where he worked in the mines and in 1968 came to Zimbabwe. First bought a restaurant to which he latter added a bakery. Moved to new premises in 1980. Experienced financial difficulties and went into partnership with an investor but retaining 51% of the stock. In 1984, the son went to South Africa to further his career but returned in 1990. There he got experience in industrial catering which he was transferring to the family business at the time of interview. The move from the old to the new premises was with the existing machines. Thus the original entrepreneur started small and the present operation took over his equipment. There was no pre-investment study. He had bought locally both old and new machines, plus some used machinery in auction. He was able to specify what he wanted and to locate some of the machines through local representatives of the suppliers.

Firm 211. Present owner's father started the business in 1950. He owned a butcher-shop and had been trained as a butcher in the United Kingdom. After migrating to Zimbabwe (via Kenya and South Africa), he worked first for the Cold Storage Commission, ran a butcher-shop and then opened a self-service supermarket (the country's first). Operated 7–8 stores in the country, but his business collapsed when the Southern Federation (Zambia, Malawi, Zimbabwe) was dissolved. His son, the present owner, went to the United Kingdom after leaving school to work and train. He obtained a three-years diploma at the Meat Institute in London, and started his own retail business (butchery), and a restaurant, in 1970. Sold both when called to the Army. In 1978 he was asked by a chain of supermarkets to establish a network of butcheries in their supermarkets, which he did. He sold half of this business to the supermarket chain in 1980 and the other half in 1989. His father, who passed away in 1989, left him the manufacturing plant that we visited and which he took over at that time. It complements the wholesale side of the business that they also operate. Thus the manufacturing operation was started small and without a feasibility study. The equipment had been selected by the original entrepreneur. These machines were collected over the years. The filling and mincing machines, manufactured in South Africa, were obtained new in 1989. For these machines they knew what to buy from previously accumulated experience, and they could interface with the supplier in specifying the equipment.

Firm 241. Individual entrepreneur (Sephardic Jew) that immigrated originally from Rhodes island, Greece, in the 1930s, started a bakery in 1953 and then a biscuit factory in 1957. Machines to make sweets were acquired in 1960–61. In 1965 he started to make chocolate. Thus the original owner started with a small operation and no feasibility study. He selected himself the equipment. The present owner, an ex-employee with same ethnic background, inherited some of the equipment (chocolate line) and acquired new machinery for sweets and biscuits manufacturing. For the recent acquisitions, there were trips abroad, examination of catalogues and observing similar machines in international fairs.

Firm 361. Enterprise started as a family business in 1947 manufacturing jelly beans and later on potato chips. It was acquired by a local brewery in 1974 and sold to two local brothers in 1979. In 1990 the enterprise was acquired by a local group controlled by the government via the Central Bank. Again this was an acquisition of a small ongoing concern. For a more recent acquisition of equipment, that took place in 1988, the Technical Director examined catalogues and travelled to Europe to decide on the purchase.

Firm 551. The company was acquired by the present owner, as an ongoing concern, in 1980. Most of the equipment was already there so the new owner played no role in selecting or specifying it. Two additional pieces of equipment were acquired recently with full specification by the present owner, based on his own experience.

Firm 591. This MNC subsidiary was started in 1959 as a distribution facility for imported goods. The first manufacturing plant was commissioned in 1961. A feasibility study was probably prepared at headquarters beforehand, and the equipment also selected by them.

Firm 1401. The original cola-based soft-drink franchise was obtained in 1949. Initial entrepreneur had immigrated from Scotland and spent a long time in gold mining. Moved to two places before getting to the present one in 1984. This was the first time that the plant was designed specifically for bottling. Short of capital, he acquired a British partner but retaining a majority of the shares. Thus the start was small, with few machines and no pre-investment study. No other information was available, but it was assumed that the original entrepreneur selected the equipment, back in 1949. More recent machinery acquisitions were done with the benefit of plant's personnel technical knowledge.

Firm 1661. Owner-entrepreneur is Ashkenazi Jew who started 51 years ago with two very old machines. Had immigrated from Lithuania to South Africa, at age 16, and worked there in mining. After a year in South Africa he came to Rhodesia. Started working for farms and selling cattle. First had an abattoir then went into sweet manufacturing and mining in 1949. Started making sweets with 18 workers (mostly females). Eight months later built a factory in present lot. Went into biscuits some 40 years ago. The original entrepreneur bought a small existing plant on the spur of the moment without previous study. For later acquisitions of machinery, suppliers from various European countries come to offer equipment.

Firm 1731. Started small, with few machines, in 1985. Owner is chef and photographer by professions. Had no formal studies. Selected himself the equipment. Had opportunity to observe similar machines locally because of his cooking experience.

Firm 71. Pig producers got together and originally approached the Cold Storage Commission to act as an agent for them. Then they organized as a cooperative in 1944. Faced with financial constraints and the need to compete, in 1993 they issued shares in the stock exchange and became a public corporation. Some of the machinery dates back to 1944.

Firm 1221. Large para-statal organization that processes most milk produced in the country. Also produces cheeses. Strong corporate culture, claims to be highly professional and efficient. Invested continuously and have modern plant facilities. Politically powerful private farmers support them. Seem to have had access to foreign exchange for investment purposes, as well as to free foreign technical assistance and training.

Firm 1351. Very small new bakery. Owner had larger frozen foods business before. He does everything with the help of three workers and his minor son. Equipment lent at no cost to him by the local Indian-Muslim community.

Firm 1951. Around 1970 six firms joined to form the current conglomerate dedicated to the production of spirits, wines and juices, with each partner contributing existing equipment.

2. Source and characteristics of equipment

It was not possible to survey every piece of machinery in each of the factories visited. Instead, respondents were asked about their three principal pieces of equipment. In some cases, more than one unit was available for each type of equipment mentioned. An average age of the equipment was estimated in such cases.

As can be seen from Table A-5, while most of the machinery is imported (from Europe) there are several cases of locally-made equipment. Among countries providing the equipment, the most frequently mentioned was Germany. South Africa was also mentioned in a few cases.

The equipment was originally bought in about equal proportions new and used. Although some relatively recent additions were reported, in a number of cases the machinery is quite old; it is quite common to find machinery 20 or more years of age in this industry. In one case the age of the equipment was not known, but it had been obtained used and reconditioned and was quite old. Equipment more than 40 years old was reported in another case.

3. Use of foreign technology

Three items are herewith considered: (i) whether the enterprise has a foreign licence or technical assistance agreement, (ii) if any help was received for commissioning the plant or installing the equipment, and (iii) if such were the case, what was involved.

As can be seen from Table A-6, five of the 13 firms in the group had licences, and six had technical assistance agreements. Three plants did not receive assistance for

Table A-5 Food and Beverages. Information about key pieces of equipment

Firm	Country	Age (years)	Bought new or used: New	Used
11	Germany,	20	2 new	1 used
	Zimbabwe	15	all new	
211	S. Africa	4	new	
	S. Africa	4	new	
	Germany	28		used
241	Germany	7–8	new	
361	S. Africa	> 20		used
551	Zimbabwe	7	new	
	Italy	> 13		used
	Zimbabwe	13	new	
591	Germany	unknown		used
	Holland	unknown		used
	Norway	unknown	new	
1401	Germany	17		used
	Germany	5	new	
	Germany	22		used
1661	Switzerland	>23	new	
	Italy	unknown	unknown	
	Germany	unknown	unknown	
	U.K, Germany	>40		used
1731	Zimbabwe	8	new	
71	Germany	0	new	
	Germany	49	new	
1221	Switzerland	4	new	
	France	4	new	
1351	South Africa	51		used
	Zimbabwe	unknown		used
1951	South Africa	>25	new	
	South Africa	18	new	

commissioning the plant or installing the equipment, and no information was available on another; all the rest received some form of outside help.

Firm 211. Received assistance from a supplier in the form of recipes. They are only required to buy the prepared sausage mixture from them; no fees or royalties are involved.

Table A-6 Food and Beverages. External technical assistance received

Firm	*Licence*		*Technical assist.*		*Outside help*	
	Yes	*No*	*Yes*	*No*	*Yes*	*No*
11		X		X		X
211		X	X			unknown
241		X		X	X	
361	X		X		X	
551	X			X	X	
591	X		X		X	
1401	X		X		X	
1661		X		X	X	
1731		X		X	X	
71		X		X		X
1221		X	X		X	
1351		X		X		X
1951	X		X		X	

Firm 241. Received assistance for the commissioning of the two biscuit production lines. Engineers from the Dutch equipment supplier came for two months the first time (1984), and for six weeks the second time (1992), to instal the machinery and train the workers.

Firm 361. They are using a foreign licence for the production of tomato-based canned products (not manufactured at the plant visited). The agreement provides for the use of the brand name and the manufacturing process. Technical assistance is provided as part of the contract. The royalty is 0.67% of sales of that line of products. They are not allowed to export to South Africa and need clearance before exporting to other African countries in the subregion. They received external help for the commissioning of new extrusion equipment used in the manufacture of potato chips. A technician from the supplier came for one week and they send their technician to his plant in France to train for two weeks.

Firm 551. Received help from the supplier for the installation of new equipment. He sent technical personnel to do the job but the workers did not receive any special training on that occasion.

Firm 591. This subsidiary of a MNC pays 5% sales royalty and gets technical assistance included in the package. While corporate policy is to grant to individual national units as much autonomy as possible, they must clear export plans with headquarters. They received help for the installation of new packaging equipment from Canadian supplier who sent personnel to do the job.

Firm 1401. As bottlers with brand soft-drinks franchises, they must buy the syrups or beverage mixtures from the franchisors; no other payments are required. They must also conform to product and process specifications and have been required by one of the franchisors to acquire old machinery (which they reconditioned) as a condition to get the franchise. For recently acquired equipment, they solved minor installation technical problems with their own personnel plus a technical expert sent by the British supplier.

Firm 1661. Received assistance from East German supplier of toffee-making machinery who sent technical personnel for the installation.

Firm 1731. Received assistance for the installation of new sachet-packaging machine. Their machine operator went to work for the supplier for six months.

Firm 71. Expect to receive Danish government technical assistance in the future.

Firm 1221. Equipment supplier helped in installing machinery and undertook to assist in maintaining the plant and training personnel. Expect to be able to get further technical assistance from sources in Europe and the United States.

Firm 1351. Through a special arrangement, the local Muslim community is supporting this small bakery.

Firm 1951. Produce some spirits under licence and receive technical services from their shareholders who are paid via dividends. Partners also provided them with some discarded equipment. They also reported receiving help from suppliers in installing equipment and training their personnel to operate it. Face export restrictions; are allowed to export but not everywhere.

C. Operating capabilities

1. General

Firm 11. Appropriate, clean, well-lighted building, but old equipment. Capacity utilization self-rated at 70%.

Firm 211. Appropriate building. Well ventilated and lighted. Some old and some new equipment. Low capacity utilization (50% according to plant manager, and 30% according to managing director). Some complete lines appeared to be idle. Seemed to be suffering from the removal of beef subsidies, and ensuing competition of imported frozen fish.

Firm 241. Adequate industrial sheds, some originally built for storage. Poorly lighted assembly sections. Slippery floors in some areas. Deficient process organization in some sections. (Plans exist for the construction of a new plant in adjacent land owned by the firm.) New and old machines alternate: those in the chocolate line are 26–28 years old, the machines in the sweets line are seven to eight years old, and the biscuit-making ovens are one and nine years old, respectively. Capacity utilization varies with demand (seasonally and by product line). The chocolate line was under-utilized. One new machine was not working because they could not programme it. On the whole, capacity utilization was self-rated at 50%.

Firm 361. Appropriate industrial sheds. Slippery floors in several sections because of spilled liquids. Mixture of new and old equipment, on average 18–20 years old. Some machines were idle because they had been replaced but not removed, others were not in production.

Firm 551. Appropriate plant, reasonably clean and with good storage facilities. Very few machines, most 15 years of age or more. Some new equipment.

Some machines were idle because of low demand. Capacity utilization self-rated at 80–85%.

Firm 591. Plant is well lighted, clean and of appropriate construction. Layout is inadequate, production is scattered in several small buildings. Substantial air-heaters and boiler plant with coal arriving to the plant's yard by special railway line. Average age of machines well beyond 20 years; boilers are over 40 years old. Packaging is mostly manual. Several lines were not working. Capacity utilization varies by product line according to demand and was approximately: 12% for condensed milk, 40% for powder-milk and 53% for cereals. Weighted average capacity utilization was estimated at 45%.

Firm 1401. Nice building with well-cared flower garden. Plant has in-situ training facilities. Serious shortage of water led to construction of water recycling plant and to digging of their own wells. Water was spilled in the floor in some sections. Equipment is mostly quite old. Do a lot of reconditioning and keep a large stock of discarded machines for future reconditioning and/or 'cannibalization' for spare parts. One line was not being used, in spite of which capacity utilization was self-rated at above 90%. (At visit time it seemed to have been substantially lower.)

Firm 1661. Production processes were scattered around and not aligned, with machines placed in various buildings, some of which inappropriate. Materials and products were scattered all over. Food materials were handled without proper protection. Dangerous working conditions in biscuit manufacturing (burnt-out ovens with protruding insulation) and in pop-corn making. A lot of the packing is done manually. Machines were well above 20 years of age; biscuit ovens are literally falling apart. Some sweet-lines machines were idle and one of the two biscuit ovens was not working. Capacity utilization was self-rated at about 60%.

Firm 1731. Small place, crowded, disorganized, untidy, not adapted to processing undertaken. Equipment is rather new (about eight years old). There was some unutilized equipment. Respondent could not provide estimate for rate of capacity utilization.

Firm 71. Very old and poorly kept plant, working one shift. Some floors were slippery and there was a lot of water in some areas. No use of gloves. Hygienic level seemed questionable but they claimed not to have had food poisoning instances. Some idle capacity in canning lines and surplus labour was apparent. Very old machinery.

Firm 1221. Very nice and well-kept plant but seemed to have redundant workers. Mostly relatively new equipment bought in the period 1985–93. Number of shifts and capacity utilization vary according to product line. Long-life milk line runs 24 hours a day at full capacity. Buttermilk line runs two shifts, and the rest of the plant, one shift. Fresh milk line could produce substantially more and was running at about 50% capacity in one shift.

Firm 1351. Small, well-organized and clean plant working one shift at an estimated 40% level of capacity utilization. Mixer is very old and the oven of poor quality. Kneading is done manually.

Firm 1951. Nice and well-kept plant. Firm provides highly subsidized housing to its workers (of which there seemed to be too many), in a village it owns next to the plant. Also have a pre-school facility for workers' children, and a canteen. Most equipment is old with a few new pieces for wine making. Plant works only one shift at an estimated 33% rate of capacity utilization. Batch production affected by high seasonal variability of demand and overcapacity in some lines which results in idle machinery.

2. Troubleshooting

Firm 11. Experienced equipment breakdowns (mixers) and problem was solved with help of local engineering firm.

Firm 211. Ammonia freezer motor burned down had to be rewound which was done by local firm, but they had to stop for a week.

Firm 241. Power failure ruined control panel and materials in process melted and had to be used as 'seconds'. Fixed with outside local help.

Firm 361. Defective control in heat exchanger led to it blowing up. Serious accident; ceiling blew off but nobody was hurt. Supplier come to fix it and a firm's technician went to South Africa to get a spare part needed.

Firm 551. Pasta machines keep breaking down because they are old. Needed replacement of gears, bushes and bearings. Local engineering firm fixed it. They also suffer of frequent vehicle breakdowns because their trucks are quite old.

Firm 591. Black production manager was reluctant to acknowledge that anything could have gone wrong.

Firm 1401. Young, dynamic plant manager, did not hesitate in reporting problems. Acknowledged having experienced serious difficulties in quality control, safety, pollution control, and other areas. The quality problem was due to poor supervision by a foreman having at the time personal problems at home. There was a serious accident with a fork lift due to poor training. Operators were sent for training and now only 'certified' drivers are allowed to operate these vehicles. The pollution problem was due to effluent with high acidity. The city now controls the effluent discharge monthly. The plant also experienced other serious problems not of a technical nature. They recently had severe shortages of water and CO2. The first due to the draught that was very severe in that area (Bulawayo) and the second because the carbon dioxide was produced from ethanol generated at a sugar refinery but the sugar crop was lost also due to the draught. They have also problems with the supply of electric power. They are subject to off-loading when consumption peaks.

Firm 1661. Owner reluctant to talk about problems. However, due to the very old age of the equipment, frequent breakdowns are to be assumed. Insulation of biscuit ovens was protruding at a few places and burns were apparent in others. Owner and plant manager only referred to problems with the electric supply. (A substation burnt down and they had to stop production for lack of power.)

Firm 1731. Had equipment breakdowns which were fixed with local external assistance. Has little in the way of equipment.

Firm 71. Experienced equipment breakdowns and bottlenecks or production imbalances. Most equipment is fixed in-house. Because of past foreign exchange restrictions, they built shop to manufacture many spare parts. Some equipment requires outside local help to be fixed. Equipment is not sufficiently flexible to accommodate desired changes in production mix. Tackled this problem by doing 'capacity stretching' via changes in the use of some equipment, and by staggering the hours of use for others.

Firm 1221. Bottlenecks were their main problem while equipment breakdowns are infrequent. Due to changes in market demand they would like to increase output of long-life milk products; not having extra capacity in that line, they have resorted to increasing the number of operating shifts.

Firm 1351. No instances of trouble-shooting were reported.

Firm 1951. Reported equipment breakdowns and quality control problems. The bottling line is too old and breaks down frequently; it is fixed by in-house personnel. Complained about low quality of locally produced wine-corks. When trade restrictions were lifted they switched to imports. Another quality problem was due to unclear bottles because of inadequate bottle-washing machinery. Also experienced difficulties with locally produced labels; use imports for their export products.

3. Quality control

Firm 11. Inspection is practical and essentially visual, both for the final product (bread) and the production process. It is important to check the moisture content in the flour which is their key raw material. They do this practically although equipment to measure the water content is available. There is no control of materials and components in process. No norms are applied, since, according to the respondent, there are no standards for bread in Zimbabwe. Samples are examined on an *ad hoc* manner during the production process. They consider their quality to be much higher than that of local competitors. Due to increased competition, and improved personnel training, the proportion of rejects (returned bread) has been reduced significantly last year (to about 5% from around 25% three and five years ago). To keep track of customers' complaints, sales people and managing director visit them; questionnaires have also been sent to retailers to elicit feedback about their product. There is no system of statistical quality control.

Firm 211. Final products (sausages), processes and raw materials are inspected practically and in a visual manner. There is no control of materials and components in process. In addition, the City Health Department conducts laboratory tests for preservative content. No norms are used. There is no routine to sample during the production process. They perceive their quality as being about the same as that of their local competition. There are no imports at this time. They claim to have a very low proportion of rejects (about 1%) and attribute those largely to insufficient rotation by the customers (i.e. products kept beyond their prescribed shelf-life). Sales representatives bring the problems raised by customers to management and weekly meetings are held to discuss them. There is no system of statistical quality control.

Firm 241. Raw materials, processes and final products (biscuits, chocolates and sweets) are the subject of visual, practical inspection. No standards are applied. There is no routine to take samples during the production process. Claim that their quality is higher than the competition for biscuits and chocolates, and about the same for sweets. They claim to have very low proportion of rejects (less than 1%). Keep track of customers' complaints by direct contact with them. Their quality control seems very primitive for a food industry establishment. Quality control is totally in the realm of production and they do not have a system of statistical quality control.

Firm 361. Have quality control of: final product (potato chips, instant coffee, cereals), production processes, raw materials and materials and components in process. Visual, practical inspection methods are employed for all four, plus laboratory tests for the first three. Production inspects quality but there is also a Quality Assurance Department which reports to the Technical Director (i.e. is independent of production). The Quality Assurance Department also supervises the laboratories. They were not using international quality norms but plan to do so, having began to instal 'quality circles' and ISO 9000 for which they received assistance from Veritas (Norway) and the Standards Association of Zimbabwe (SAZ). They are subject to health regulations because they produce food products. Routines exist to take samples during the production processes; some are taken hourly and others every two hours. There is one quality controller plus five quality inspectors, and their manager, in the plant. Laboratories conduct various specialized tests, for example, protein test for baby food and high-fibre cereals. They claim that their potato chips are of much higher quality than the competition's (both local and imports), and that their coffee is of higher quality than the local competition, but admit that their corn flakes are of lower quality than competing imports. No statistics were available about rates of rejection, but the respondent claimed that they had been improving over time. A public relations officer receives all complaints from the customers, assuages them, and then channels the complaint to Quality Assurance and Production. Quality Assurance (under the Technical Director) then prepares a written report acknowledging or disputing the facts. This report is circulated within the firm and remedial action is taken if necessary. The customer is then approached with an answer. There is no statistical system of quality control.

Firm 551. Final product, processes and raw materials subject to visual, practical inspection methods, but final product is also tested in outside laboratories and SAZ because of health regulations applying to food products. They also do some testing to measure PH and sterilization, for spices and sauces. They claim that their spices and herbs are of higher quality than the competition's. They professed to have had a rejection rate of about 3% last year, mostly attributed to breakage in shelves and poor packaging materials. Claim a quality improving trend due to improvements introduced in the packaging (better bottles) and better cleaning of organic products (spices and herbs). Customer complaints are received through the mail and investigated in the factory. Do not have a statistical quality control (SQC) system.

Firm 591. This subsidiary of a MNC has quality control in production and independent of it (similar to Firm 361). They carry out visual, practical inspection in all phases: final product, production processes, raw materials purchases, and materials in process. In addition, laboratory tests are done for chemical, physical, and biological characteristics. Savory testing is also part of their quality control. Professed to apply 'total quality management' (TQM). Although contemplating the introduction of ISO 9000, they claimed that their system has stricter requirements. In addition to meeting HQ requirements they are subject to controls by the Ministry of Health whose personnel take samples for testing. Export shipments are checked by SGS. They have a routine to take samples for testing during the production process. They claim not to have local competition, but the Dairy Marketing Board (DMB) is actually competing with their dairy products. In fact, the newly appointed plant manager was in the process of trying to establish independent quality assurance procedures for future supply of fresh milk that he was trying to organize directly with dairy producing farmers. They claimed to have very low rejects level, less than 1%, and that includes products expired (i.e. beyond their shelf-life). They keep track of customers' complaints via the sales force. Quality assurance does the checking, with inputs from production, and the sales force deals with the consumer. They do not have a SQC system, but they graph process pressure and temperature and these values are checked hourly.

Firm 1401. They have both visual practical inspection and laboratory tests for product, processes, raw-material purchases and materials in process. Their main raw materials, water, sugar, syrups and carbon dioxide, are all subject to various laboratory tests. The final product is checked at random in the warehouses, sales trucks, loading docks, and by the merchandiser in the retail outlets. They participate in periodic (every 3 months) technical meetings among the five bottlers operating in the country, at each other's site, where quality control is also discussed and test results are compared. This permits them to have some control on the calibration of their instruments. These meetings were so successful that they have been extended to the plant supervision level. They must of course meet the technical specifications set up by their international franchisors as well as local health requirements. There is an established routine for inspection of samples taken during the production process. They consider their quality to be similar to that of local competitors. (The comparison was probably meant with bottlers in other locations since regions are assigned and bottlers do not seem to compete within each region.) They have a low proportion of rejects (less than 1%). They receive customer feedback via the sales force and by direct contact through the merchandiser, who checks shelf-life and storage conditions as they apply to their products.

Firm 1661. Inspection in all phases (product, process, raw materials and materials in process) is visual and practical. Weights of sugar, glucose, etc.,and boiling temperature (95 degrees C due to the altitude) are checked because they are important in the process. Final product is checked together with the package. Local health department sends inspectors periodically to retrieve samples for testing in their laboratories. No standards are applied, and there is no routine to take samples during the process. The plant manager admitted that the quality of their biscuits was lower than that of the local competition, but considered the quality of their other products to be similar to that of other producers in the country. Rejects are less than 1%, and they are sold as 'seconds' at give-away prices. Customer complaints are handled by direct contact with the client or via the sales force. Their motto is that the customer is always right and they give compensating credit without arguing. No SQC system.

Firm 1731. They inspect the final product and the materials in process by visual, practical means. Some products are tasted. Apply no standards but are contemplating to use EEC standards for future exports of canned products. There is no routine for sampling during the process. The respondent was unaware about how their quality rated in comparison with other local or foreign producers. Last year they had a reject rate of about 4%. Mostly pickled products returned by supermarkets. They attribute it to poor packaging. Complaints are transmitted via the sales force and also by direct contact. No SQC system.

Firm 71. Claimed to apply laboratory tests and standards to the final product, production processes, raw materials and other purchases, and materials and components in process. They also take samples during the production process, but there is no SQC system. Testing takes place daily and they try to comply with national and international standards such as ISO 9000 and EEC standards. Have their own microbiology laboratory. Claimed that their quality compares favourably with that of locally available similar products, and is equal to that of imports. Rejects were quite low, less than 1%, and were mostly problems of taste or colour due to old, inadequate machinery. All rejects go to animal feed. Returns were mostly due to bad handling after leaving the plant. Merchandising agents deal with clients' complaints.

Firm 1221. Laboratory tests and standards were used to inspect: final product, processes, raw materials and other purchases, and work in progress. Claimed to apply national and international standards. Had a sampling routine for inspection during the production process but no SQC system. Claimed to have the same quality as imports. Prior year's internal rejects rate was less than 0.5%. Some quality problems were due to the lack of a foolproof system to wash plastic containers before filling them. Most other problems arose because of poor cooling facilities after the milk is delivered to the stores. They keep in direct contact with customers to track down and answer their complaints.

Firm 1351. Inspection of final product, production process, and purchases, is visual and practical. No norms are followed, and there is no sampling inspection routine during production. No system for SQC. Rejects were less than 1%. They maintain direct contact with customers to keep track and satisfy their complaints.

Firm 1951. Conduct laboratory tests and apply standards to inspect the final product as well as raw materials and other purchases, and visual, practical, methods to inspect processes and work in process. Use national and international norms prescribed by the Food and Beverages Association of Zimbabwe. They also have a routine to inspect samples during production, but no SQC system. Their quality was equal to that of others for locally produced wines, while it is similar to that of imports for spirits (they face no competition in those spirits produced under licence). They had less than 1% internal rejects. Most recent problems had been with wine and quite infrequent since the cork difficulty had been taken care of. Keep track of consumer complaints through their sales force and sales outlets.

4. Maintenance and repair

Firm 11. They have repair and maintenance (R&M) staff (officer in charge plus two people). They do not have a preventive maintenance system (PMS), but the workshop manager checks the machines weekly. Oiling is also done every week and spare parts are replaced when needed. Outside help is required for 'major' work such as replacing gears. They do not experience frequent breakdowns.

Firm 211. Have R&M staff (manager and two people). They had a system for partial preventive maintenance; it included scheduled maintenance of blades, and other machine parts as well as defreezing. Every three weeks they do oiling and greasing. Defreezing is done very day for one hour. Blades are resharpened or changed every week. Other parts are replaced as needed. They use R&M outside help for refrigeration and electrical problems. Equipment breakdowns are not very frequent; maybe once a week.

Firm 241. They had R&M staff reporting to the Managing Director (MD) and including: one senior mechanic, one electrician, two journeymen and 10 assistants. No PMS, but each year they close down for three weeks, around Christmas, for an overhaul (for example, to change cooling belts in the sweets-making machinery). Oiling is done weekly, every Friday. No frequency for replacement of parts has been established. Outside help is used when required, e.g. for compressors and boilers. Equipment breakdowns are frequent but generally minor in character. They occur mostly in the sweets plant, particularly during humid weather which causes stickiness.

Firm 361. Had in-house R&M staff that includes a Plant Engineer plus 10 people, and were planning to start a preventive maintenance system (PMS) next year. There is no formal system to check the machines, but foremen were supposed to do it weekly. Oiling, etc. is also done once a week and there is no established frequency for parts replacement. Outside assistance is required for problems with: refrigeration, compressors and the gas plant. Breakdowns are common because some machines are quite old and tend to break down more frequently.

Firm 551. They had no R&M staff and used local engineering firms as the need arose. There was no PMS, and they checked the machines only when they broke down. Oiling was done every week; there was no established frequency for parts replacement. Breakdowns were quite frequent (weekly) and concerned mainly the hammer-mill and the pasta-making machine. R&M activities are under the direct responsibility of the factory manager.

Firm 591. Had a Plant Engineer responsible of all plant facilities, and including R&M. Besides a draftsman, and an assistant engineer, there is a workshop foreman with 13 employees, two electricians, one senior boiler operator and eight employees, one staff in charge of the air-heater and three people in the technical stores. They had a system (used in the MNC to which they belong) of 'total preventive maintenance' (TPM) but it was not very effective. Thus they divided the plant equipment into three categories: critical, semi-critical, and non-critical. In the first category fell, for example, the boiler, air-heaters and the tower roller drier. These undergo yearly overhauls. For the second category they keep an appropriate stock of spares, and for the last, there are no precautions as they are easily repaired. Oiling and other required servicing is done daily. They have a quite complete in-house R&M service but require some external assistance for specialized services. They may send out scales for calibration, parts that need balancing, or hire specialized refrigeration services. Data on frequency of breakdowns was not available, but being their machines quite old (boilers more than 40 years) it is fair to assume that breakdowns are relatively frequent.

Firm 1410. Had R&M personnel (reporting to the production manager), but no established routine for preventive maintenance. They used to have annual machinery overhauls but had abandoned the practice. Now they look for weak spots and organize the maintenance program around them. Greasing and oiling is done daily, and other servicing once a week. There is no established frequency for parts replacement. Outside services are used for refrigeration repairs. They experience frequent, but minor, breakdowns. Machines are quite old, and require attention. Have done a lot of reconditioning of old equipment.

Firm 1661. They had a R&M shop, which also does repair work for mines owned by the entrepreneur, with a mechanic in charge plus 10 staff, three of which are qualified as journeymen. There is no preventive maintenance routine but every year they have a close-down period at which they service one of the two boilers. Every Friday machines are steamed, greased, oiled, etc. There is no established routine for parts replacement. Local engineering services are hired when the workshop is very busy and for specialized services. They claimed not to know about frequency of breakdowns, but in view of the age of the equipment it can be assumed that they were frequent.

Firm 1731. No in-house capability, and no established routine for R&M. Machines are serviced weekly. The factory foreman is responsible for R&M and the machine operators are expected to do the servicing, i.e. oiling, replacement of belts, etc. Outside help is required for all other work. Breakdowns are frequent but no further information was available.

Firm 71. Possessed a R&M shop but had not yet established a routine for preventive maintenance of the plant's equipment. Carried out daily checks depending on the number of hours the machine has been operating. Breakdowns were frequent, even daily. Used no outside help as they were able to strip most machines to replace standard parts or get spares made.

Firm 1221. Maintenance department was in charge and there was an established routine for preventive maintenance whereby they stopped to check machines at established intervals whether they presented a problem or not. Main pieces of equipment are subjected to yearly overhaul. For this purpose outside specialists are brought from Europe or South Africa. Claimed that breakdowns were infrequent and that they had the capacity to strip most machines to replace standard parts or to get spares made.

Firm 1351. Owner of this very small bakery did whatever R&M was needed. Outside help was obtained when required.

Firm 1951. Production management was in charge of R&M and they had a shop for that purpose. Main problems had been with bottling line which was quite old and had continuous maintenance requirements. Also with the water supply that caused deposits that tended to clog the pipes. Two fitters were continuously assigned to that line. Otherwise they claimed to experience only minor and infrequent breakdowns. Claim to have a complete in-house R&M capability. Once a year the still is shutdown for a complete overhaul. Other servicing, such as oiling, etc., was done on a monthly basis.

5. Industrial engineering

Firm 11. No continuous organized effort to improve productivity. No time-and-motion studies are carried out. No production standards exist. They claimed that their 'efficiency' was about 75% of the industry's norm. Production is planned day to day on the basis of sales orders brought by sales personnel, and the cyclists and drivers making deliveries. Given the nature of the product (bread) they maintained inventory control only for expensive raw material ingredients kept in the store room. Thus there was neither work-in-progress nor final product inventory. No computers were used on the above work. Layout changes were carried out on occasion of expansion or equipment replacement. Last change was done four years ago, when the dispatch area was enlarged. They took the opportunity to improve the layout to avoid unnecessary work duplications.

Firm 211. No continuous, organized, effort to improve productivity, and no time-and-motion studies are carried out. While there were no production standards, they operated with a monthly 'budget' and gave an incentive bonus based on meeting the overall budget plus individual performance. The bonus could be up to 18% of the monthly salary. They are unaware of how their productivity compares with that of other local or foreign firms. Due to the nature of the product (sausages) they must schedule production on a daily basis based on orders obtained by the sales representatives. Inventory control was geared to production process control and sales. Stock-taking of work-in-progress was done weekly and monthly. The data was recorded manually and later entered to the computer. Layout is re-examined on occasion of plant expansion or replacement of equipment. Last change was done a week before the interview to accommodate a new packaging line.

Firm 241. No organized effort to improve productivity existed, but they were planning to introduce an integrated system of production and cost control to be carried out by a recently hired MBA (majored in accounting). No time-and-motion studies done. Planned to use historical production data to develop an incentive system. Owner (and Managing Director) presumed to have inside information about productivity in both large and small local competing plants, and claimed that that of his plant was higher. Had no information on foreign productivity but presumed it must be higher, particularly due to poor packaging capabilities in Zimbabwe. Scheduling leads varied from only two to three days in advance for certain perishable products with large demand variations, to two weeks in advance for stock and export orders. They kept control of inventories of raw materials and final product. Stock was taken every week. Work was done manually; available computer was used for payroll, profit and loss accounting and correspondence. Layout was re-examined only on occasion of plant expansion or replacement of machinery. Last change was done when they bought new machines for sweets and biscuits, and the next one will be on occasion of installing a chewing-gum production line.

Firm 361. No effort to improve productivity but the respondent (Technical Director) recognized the need for it. They did not carry out time-and-motion studies at interview time. Had a work-study engineer, trained abroad, but he left for another job in the group and had not been replaced. Production standards were set historically by work-study, but nobody was responsible for that task any more. Respondent estimated that their productivity was about 20% lower than that of similar installations in South Africa and Europe that he had visited, and attributed it to the lack of local process engineers. Production planning varied with the product. They make weekly plans, except for coffee which requires up to two weeks lead. Their inventories of raw materials and final product are under control. At interview time they did not need more than three months stocks for raw materials, and their work in progress was only for about two days of output. Computers were used for production planning but not for inventory control. Some of their PCs were integrated in a network; others were not. Plant layout was only re-evaluated when expansions or equipment replacement takes place, but respondent recognized the need to assess it on a continuous basis. Last change was done on occasion of installing a new extruding machine and the work was done in-house.

Firm 551. No organized activity to improve plant productivity. Production standards had been set by previous production manager. No time-and-motion studies carried out. Unaware about how their productivity compares with that of other similar plants. Production was scheduled on the basis of orders, generally for a few days in advance, except for rush orders for spices and pasta that could be executed on a day-to-day basis. Exports took longer planning, at least one month ahead. Claimed to keep control of inventories of raw materials, work in progress and final product. Took stock every two weeks and had daily product checks. Used a card system and had computerized it but it was not working at interview time. Used an IBM-compatible PC with a manufacturing and accounting system software. Layout was only re-examined when changes in the plant were carried out. Last change was done six years ago; it involved creating extra office space and changing the hammer-milling machine.

Firm 591. No organized effort to improve productivity but were reviewing production processes with a view to introduce improvements. Production standards were set by a committee on the basis of previous experience. A Christmas bonus plus profit-sharing scheme was applied to the whole plant, but it was not tied to individual performance. Their local, large, competition for powdered milk was the Dairy Marketing Board (DMB). Claimed that productivity comparisons are difficult due to low capacity utilization in their plant (*sic*). Estimated that it was the same as in other similar plants in other African countries of the MNC they were part of, and lower than that in South Africa. They prepared annual programme and budget. Also did quarterly and monthly updates based on weekly sales. For exports they worked by order, but there was always some advance planning, except for Zambian importers who brought cash with them and came to the door to pick up the merchandise. They controlled inventories of raw materials and final product. When a requisition comes in, they key in the data in the computer to verify if the required stocks are available. They used a TETRA 2000 system plus some special HQ-provided software. All together had 12 PCs, but the planning of production and inventory control were not fully integrated yet. PCs are also used for accounting, payroll, sales analysis, etc. Plant layout was reassessed when new equipment was to be installed. Last change was done apropos of installing a new packaging machine to automate part of the process which is however still done manually to a large extent.

Firm 1401. No organized effort to improve productivity. Production standards were machine-determined. They knew, from knowledge of other national bottlers, that their productivity was about the same. Unaware about foreign plants. They scheduled production day to day. Franchised soft-drinks were fast-moving products. Production programme was based on a monthly sales budget. They controlled inventories of raw materials and finished products, trying to keep minimum stocks for lack of space. Finished products were stored no longer than four days; raw materials, like sugar, for up to 10 days, carbon dioxide for up to two weeks, coal for two weeks. Had two PCs which were used for production planning and inventory control. Plant layout was re-examined only when installing new machines. Last modification had been made 18 months ago.

Firm 1661. No formal effort to improve plant productivity. They set standards of production based on experience. No work study was done. Have a bonus system paid every half year and not tied to production. Claimed to be unaware of the productivity of other local or foreign plants in the industry. They planned production for stock and orders when the stock of a certain product is depleted. Stocks were checked every week and they also prepared weekly production plans. Kept control with a cardex system, and computer, of raw materials and finished products stocks. Took inventory every month. Layout was checked when new equipment was brought in. ...'There is room, and we are quite satisfied with the layout'. (However, see above the contradictory assessment of the plant's layout made by the survey staff in section Operating Capability. General.)

Firm 1731. No organized productivity improvement effort. No production standards. Did not know about productivity in other plants. Production was planned according to orders a few days in advance. No formal system of inventory control existed (neither for raw materials, work in progress or final product). Layout was not assessed and there had been no recent changes. (There was little machinery.)

Firm 71. There was no organized effort to improve productivity. Merely struggling to keep the equipment running. Had a monopoly before the opening up of the economy. Did not carry out time-and-motion studies, but claimed to study the machines' capacity and knowing the average labour productivity the factory manager was able to establish production standards. Did not know how their productivity compared with that of other suppliers, but assumed that it was lower than that of foreign suppliers. Scheduled production in advance but the lead times varied. Some orders are met on a day-to-day basis while others are part of a weekly programme to meet production or stock targets. There was also some longer-term planning, at least a month ahead, for certain internal or export markets. Kept inventory control only for final product. Claimed to be in the process of computerizing the stock control system. Did not examine the layout situation periodically, but were engaged in a large building expansion. Expected that their new plant would have a good layout and would meet EEC temperature control standards.

Firm 1221. There was no formal effort to improve productivity at the *plant* level; claimed, however, that the *firm* was engaged in such an effort. Did not carry out time-and-motion studies and did not have production standards, for example for manual packaging. Did not know how their productivity compared with that of other local plants; assumed that it was lower than that of foreign suppliers. Advance, long-term, scheduling of production was carried out at the firm level with the plant responding to it. Claimed to have inventory control for raw materials, work-in-progress and final products. Plant layout was only examined on occasion of an expansion or replacement of equipment.

Firm 1351. No productivity improvement studies and no time-and-motion studies. No setting of production standards. Unaware how its productivity compared with that of other bakeries, local or foreign. Production was planned on a day-to-day basis. A cardex system was in use to keep control of raw materials and final product. There was no use of computers and no layout analysis.

Firm 1951. No formal efforts to improve plant productivity and no time-and-motion studies were carried out, although they were aware of the need. No setting of production standards. There was no other local producer of spirits. In wine they did not know how they fared *vis-à-vis* other large local producers. Assumed that their productivity was lower than that of foreign producers. Production was planned at least one month ahead. Carried formal inventory control of raw materials and final product. No computers were used for those tasks. Plant layout was continuously being assessed, and they recognized the need for improving it.

6. Industrial safety and pollution control

Firm 11. Responsibility for safety or pollution control is not assigned to any office or person. They claimed to have written rules about safety but not for environmental control. Fire Department conducts fire drills at the plant and they have staff for emergency and first aid. Presumed that their safety record and procedures were about the same as those in the rest of the industry, but believed that their pollution control record was better.

Firm 211. Nobody had responsibility for industrial safety and pollution control. But they claimed to have had very few accidents and that they trained their workers well. No written rules for either safety or environmental control exist. They did not conduct fire drills, but serviced their fire extinguisher twice a year. They rated themselves at about the same level than the rest of the industry in safety matters, and below it in pollution control.

Firm 241. The responsibility for safety was not assigned to any office or person. Nurse visited the plant twice a week. Employees were sent to first-aid Red Cross seminars, and they claimed to have a good record with the Workers' Compensation Fund. Responsibility for pollution control is also not assigned; their main problems were with the fumes emanating from the boilers, and the dust arising from wood-working. No safety or pollution rules. They claimed to carry out fire drills, and considered their safety record to be better than that of the rest of the industry. On environmental control they felt that they were at about the same level as the rest of the industry.

Firm 361. There was a 'risk control' officer reporting to the Managing Director, plus a Safety Committee that included the group engineer, a visiting doctor and a nurse from the clinic. An environmental policy was in preparation by a task force under the direction of the Technical Director. Members of the task force were: Group engineer, head of quality assurance, risk controller, and representative from procurement. The policy being prepared was for the whole Group and not just for the plant visited. They claimed to have written rules for safety included in the 'Employee Handbook'. Fire evacuation drills were done but not with sufficient frequency according to the respondent. They had well-marked exit points and had trained a safety officer. Also conducted first-aid briefing sessions. They believed that both their safety and environmental control records were better than those of the rest of the industry. Had some problem with effluent discharge, but it was not serious. With technical help from Denmark, they were using sucking tanks and planning to use the waste residual to produce bio-gas.

Firm 551. Factory's supervisor instructed his workers on safety procedures. Nobody was responsible for pollution control. There were no written rules for either safety or environmental control. Foremen were trained to use the fire extinguisher, and a representative of the Workers' Compensation Fund came to talk to the workers. They rated their safety and pollution control record and procedures as better than those of the rest of the industry. Interviewers noted that the hammer-mill had no extraction system.

Firm 591. A supervisor, under the production manager, was responsible for safety. For pollution control, there was a committee of five people chaired by a supervisor. Claimed to have written safety rules. Their Worker' Compensation insurance rate had been reduced due to a low number of accidents during the previous two years. No rules existed for environmental control. The City Health Department checked the plant's effluent. They claimed that local legislation is however less stringent than the MNC's (they were part of) own internal rules, but no policy has yet been developed for the Zimbabwe plant. They conducted periodic fire drills and had designated a fire-fighting team. They felt that their record and procedures were better than those of the rest of the industry in both safety and pollution control. (Interviewers had some doubt about the accuracy of this assertion as it applied to environmental control.)

Firm 1401. They had a safety committee with a 'worker' in charge. Nobody had been assigned responsibility for pollution control, but they check boilers' effluent. They claimed to have safety instructions for the use of safety equipment and also for the use of fork-lifts (had had a serious accident). No rules existed for environmental control. Said to conduct periodic fire drills and their fire teams had been trained by the local fire brigade. Had no idea about how their safety and environmental control record and procedures compared with those of the rest of the industry.

Firm 1661. Claim that the Plant Manager was responsible for safety, and that the City's Safety Board sent materials about the subject on several occasions. Responsibility for pollution control had not been assigned to anybody. No written safety or pollution control rules existed. Claimed municipal control of their effluent. They did not conduct fire drills but had fire extinguishers and the Fire Department trained their workers. Felt that their records and procedures for safety and pollution control were about the same as those of the rest of the industry. (The interviewing team had doubts about the accuracy of this self-assessment.)

Firm 1731. No written rules and nobody was assigned responsibility for either safety or pollution control. No fire drills were carried out. Respondent could not compare their procedures and records on either area with those of the rest of the industry. One pollution control problem concerned the smoking of meats, a process for which there was no extraction system.

Firm 71. Claimed to have assigned responsibility for industrial safety to a Safety Committee comprised by management and workers' representatives. Nobody was in charge of pollution control. Seemed to think that it was not an issue in Zimbabwe. Claimed to have written rules on safety but not on environmental control. No fire drills were carried out. Maintained their fire extinguishers annually. Claimed to have better safety records and procedures than other firms in the industry, and to be at about the same level as the rest of the industry in pollution control matters.

Firm 1221. Safety responsibility was entrusted to a committee. Nobody was responsible for environmental control. Claimed to have written safety rules, but no rules for pollution control. No fire drills were carried out. Believed that their safety records and procedures were better than those of other firms in the industry. (They really enjoy a market monopoly situation.)

Firm 1351. No responsibility had been assigned for either safety or pollution control. No written rules existed for either safety or environmental control. Had been instructed on basic fire evacuation procedures. Considered their safety record to be better than that of other firms in the industry and claimed to experience no pollution problems.

Firm 1951. Responsibility had been assigned for pollution control, but not for safety. Claimed to have a written safety code of conduct, but no rules for environmental control. Fire-fighting drills were carried out often, and workers receive protective clothing. Rated their safety records and procedures higher than those of other firms in the industry. Similarly for pollution control. Had a good water-wastes purification system, and claimed that after its installation fish returned to the lake receiving the effluent.

D. Technological functions and organization

1. General

Firm 11. Recipes are collected from various sources, including cooking books. They also had their own recipes. No technical offices existed in the plant, and no other documentation was on file.

Firm 211. Recipes originated with father of present owner. Recipes are written down. Some information on spices was obtained from South Africa. There were no technical offices and no other documentation was on file.

Firm 241. Chocolate formulae, as well as other recipes used in production were written up. Many were inherited from the original entrepreneur and owner. No technical offices, and no other documentation was on file.

Firm 361. Each product had its own manual, including product specifications, process flow-chart, quality inspections, and a check-list for production personnel. Their technical offices included: quality control laboratory (with three chemists and analysts and three laboratory assistants), and research and development unit, which included a pilot plant, with seven employees (one manager, three food technologists, one technician and two research assistants).

Firm 551. Had written recipes for the cooking and other food preparations, allegedly made up by the owner. No technical offices and no other documentation was on file.

Firm 591. Manuals with manufacturing instructions were available containing product and process specifications. Also laboratory standards and specifications. The quality control laboratory has 15 employees. R&D was not carried out in installations separate from those of production. However they had developed some local products (cereal and milk-powder infant formula) which were subject to HQ approval (granted). They also had a strong plant engineering department with 31 staff!

Firm 1401. Brand-name soft-drinks franchisers had detailed product and process specifications, and they were not allowed to introduce any changes, even very small ones. Laboratory testing specifications were also available. The quality control laboratory employed eight people, most with practical training only. They also controlled the water treatment plant.

Firm 1661. Written recipes were on file with the plant manager. According to the plant manager, workers received written manufacturing instructions. The R&M workshop prepared drawings when they manufactured parts, but those were not necessarily for the manufacturing plant as they also served the mines owned by the entrepreneur. No other technical offices existed.

Firm 1731. Recipes were only known to the managing director. No technical offices and no documentation on file.

Firm 71. Kept written recipes and production manuals. Had quality control laboratory with one employee. Claimed to also experiment for product development. No other technical offices, documentation, or files.

Firm 1221. Technology was largely codified in manuals and blueprints. Claimed to have quality control laboratory and R&D office for the whole firm, but did not know the number of staff assigned. There was good technical documentation, and a number of technical offices, but at the firm level.

Firm 1351. Developed their own bread preparation formulae. Recipes for the various products were written up and the workers were aware of it. No technical offices, and no other documentation or files.

Firm 1951. Technology for the production of both wines and spirits was codified. They had one employee assigned to product and process engineering. Claimed to also carry out R&D, but not in an 'institutionalized' way.

2. Product design and product change

Firm 11. Wife produced home recipe for biscuits which they then adapted for industrial production, otherwise little or no changes to main products: bread and confectionery. The minor changes made were to suit local tastes and purchasing power. Last year introduced 200 grams biscuit package containing 12 pieces, and three years ago, industrial catering as a new product (service) offering. Of course, these changes were carried out in house.

Firm 211. Claimed that their products had been self-generated according to their own specifications. They have done some, but not very significant product changes, in particular to lower the meat content and thus price of certain products. No new product had been introduced the year before, but about two years before they had introduced french and garlic baloney, cooked tongues and roasted beef. For the baloney they used recipe handed over by father (original entrepreneur). For new, 'extended' products, they plan to get technical assistance from South Africa. They intended to introduce vacuum-packaging for chicken cuts (now prepared for third party exporter), and smoking of meat, trout, etc.

Firm 241. Products are either self-generated, to domestic specification, or copy of similar, imported product. 'Nick bar', a new type of chocolate bar prepared without cocoa butter, is example of copy of South African product. This product was introduced to lower the cost to better accommodate to local purchasing power. Introduced new sweets and chocolate products this year. They have also introduced new products in the last five years. All products are developed in-house without outside help.

Firm 361. They continuously modify existing products and also introduce new ones. The latter are all based on their own, self-generated, domestic specifications. This work was done under the supervision of the Technical Director's Office. Once a new product is developed, a manual for it is prepared. During the year prior to the interview, new products had contributed Z$ 8 million (some US$ 1.2 million) to revenue. The most recent product innovation was a new high-fibre cereal of the bran flakes type, that was an improvement on an existing one. They are now in the process of developing a meat product with an 'extensor' to lower its price. All the work of product development and modification was done in-house, but they sometimes required outside help for the testing of samples. For example, the new high-fibre cereal had to be tested in a South African laboratory. Their R&D Director was in touch with a number of research institutes. It is worthwhile noting that this firm had a specific set of objectives for their R&D work. Their expectations were that R&D will contribute to turnover by introducing new, marketable products, and that it will also result in cost reductions sufficient to meet R&D expenditures. In recent years R&D contributed (with new products) to turnover in the following proportions: 1988, 1.9%; 1989, 0.95%; 1990, 1.6%, and 1991, 1.6%.

Firm 551. Developed their products by themselves and according to their own specifications; examples: mixed peppercorn spices, hot chilli. The last product, introduced some six months before was a lemon grass power herb. Two years before they had prepared an orange marmalade with their own recipe to utilize surplus oranges in owner's farm. All the work was done in the plant but the product was sent outside for microbiological testing.

Firm 591. All branded products were made according to foreign HQ specification. Other products were locally developed to domestic specifications. They have introduced some, but not very significant, product modifications. They have developed products to use local materials, for example, soya and sorghum, as well as to take into account local purchasing power. Last year they introduced two new products: an infant formula, and 'tea-cup' milk powder. New products must be approved by Directors in Switzerland who were pleased with their quality.

Firm 1401. Products were branded and foreign specified. Although they know for a fact, that the gas (CO_2) content was lower in Italy for one of their international brand franchise soft-drinks, and in Zambia for another, they were not allowed to introduce any change. They would like to be able to use less carbon dioxide because a high gas content increases equipment wear and tear. The year before they had introduced two new soft-drink products: one with pine nut and cherry plum, and the other with grape flavours. They were mainly adopted from South Africa. Thus they were only using formulations obtained from outside sources.

Firm 1661. New products were generally modified versions of existing ones. Recently they had introduced modified versions to differentiate some product; also to meet new market needs (examples were: new toffees, marshmallows, 'Jurassic Park' dinos). These new products were originally prepared by the Plant Manager (technically trained in Germany and with long experience in the industry) who tried out their manufacture in the production line before going into full-scale production.

Firm 1731. They have not modified or introduced new products. However, they were planning a new plant for the canning of fruits and vegetables.

Firm 71. Main product design was self-generated according to domestic specifications. Claimed to have introduced a number of new products, i.e. a variety of sausages. The product changes introduced were not very significant and were done to suit local tastes, use locally available materials, and to make use of existing equipment. They have also tried to differentiate their product and to meet other market needs. However, no new products had been launched in the prior five years.

Firm 1221. Claimed to have in production some products copied from recently developed ones in the US and Europe. Have introduced some, but not very significant, changes, mostly in packaging to differentiate the product. Had not introduced any new product the previous year but some in the last five years. The latter referred to long-life milk products which were designed with outside help. Such help was also utilized for the testing of samples.

Firm 1351. Product design was self-generated. Changes introduced were not very significant, e.g. use of vegetable, instead of animal, fats, and the use of vegetable oils. These changes, developed in-house, were done to accommodate religiously-based consumer preferences. They had been carried out within the last year as the business is quite new.

Firm 1951. All their products were originally of foreign design. They had introduced some, not very significant, changes involving labelling and packaging for product differentiation reasons. They must compete in particular with South African products. Had not introduced any product in the last year, but some within the last five years. An Australian enologist, who stayed with them for three years, assisted them to bring their wines up to good quality standards.

3. Process changes

Firm 11. No changes made in the production process. They thought at some point about purchasing an automated bakery plant but gave up the idea because it would have been very expensive for them.

Firm 211. Minor product improvements which required some process or recipe modification were carried out in the plant.

Firm 241. Some product changes (e.g. chocolate without cocoa butter) required process modifications. Generally, no significant process changes were carried out.

Firm 361. They continuously improved production processes. Two such recent changes were: (i) in the manufacture of corn-flakes they installed a pre-warming device to improve the drying; it resulted in 18% more output and a cost reduction. Was carried out by a food technologist under the direction of the R&D Manager and the Technical Manager. Implementation took four months. (ii) The production of ice cream cones was improved by introducing a new ingredient. This resulted in less breakage. The change was carried out by R&D personnel.

Firm 551. No significant process changes.

Firm 591. No significant process changes introduced. They were working on energy saving.

Firm 1401. Were working to reduce energy consumption. Planned to instal a power factor control system to reduce consumption at peak loads. It was expected that their monthly electricity bill would be reduced by 50% from approximately Z$ 26,000 (US$ 4,000) to Z$ 13,000 (US$ 2,000). The cost of the new equipment and installation was about Z$ 100,000 (US$ 15,385); i.e. the investment would be recovered in about eight months. Other planned changes also had to do with plant facilities (recycling system for water saving), and equipment (reconditioning and updating of equipment) rather than with process improvements narrowly defined.

Firm 1661. No process changes; they only modified and introduced new products.

Firm 1731. No process changes.

Firm 71. Claimed to have carried out technical changes in energy saving and capacity-stretching. They were conducting an energy audit for which they had contracted, at low cost, local technical personnel for two months. Had introduced staggered working hours to balance and increase capacity. They recognized the need to upgrade the skills available to them for technological improvements.

Firm 1221. To stretch capacity and cope with bottlenecks they had introduced multiple (two and three) shifts in some sections of the plant.

Firm 1351. No process changes had been carried out.

Firm 1951. With outside help, had introduced power rationing to lower their energy costs. It had not been expensive and they were satisfied with the results. They had also increased available storage capacity within the existing building, using their own personnel and at low cost. Results had been as expected.

4. Use of technical support services

Firm 11. Use R&M services of local engineering company, twice or thrice a year. Only other services in which they would be interested were: personnel training and labour relations. They complained about thefts (by delivery drivers) and facing great difficulty in firing an employee even for a serious reason. Felt that lack of education, and past political indoctrination, negatively affect workers' performance.

Firm 211. Used outside services for quality control (QC) and personnel training. The QC was free and done by the City Health Department. They have sent middle management to local training courses in management. The trainers had a US franchise. The course lasted six months and met weekly. Both the Managing Director and the Plant Manager, were asked about the technical support services that would be beneficial to them, and their ranks were averaged. The result was: (1) QC testing, (2) trouble-shooting technical assistance, (3) process improvement, (4) productivity improvement techniques, and (5) repair and maintenance.

Firm 241. Only used yearly R&M service for boiler, done by a local company. The owner-MD ranked the support services deemed of potential benefit for the firm as follows: (1) productivity improvement techniques, (2) personnel training, (3) labour relations, (4) process improvement, and (5) energy saving.

Firm 361. Used various types of external services, both local and foreign. For corroboration of tests they used the SAZ as well as various government laboratories. They have also used Veritas for installation of ISO 9000. They employed R&M services for compressors, gas plant and refrigeration facilities. They have also utilized local providers of management and computer training courses. Finally, for brand products produced with a licence, they received technical assistance from the licensors. As to potentially beneficial technical support services, they were ranked as follows: (1) productivity improvement techniques, (2) process improvement, (3) energy saving, (4) pollution control, and (5) product design.

Firm 551. Used quality control and R&M outside services. The quality tests were done by SAZ and local biological testing laboratories. Have used a local engineering firm to fix the pasta machines. Such services were frequently required as the machines are quite old. Their ranking of desirable technical support services was: (1) personnel training, (2) productivity improvement techniques, (3) quality control testing, (4) industrial safety, and (5) labour relations. They also expressed interest in calibration of control instruments (for scales, thermometers and PH meters).

Firm 591. They have used the following outside services: personnel training, conducted at Headquarters in Switzerland as well as in Harare Polytechnic; R&M for specialized machinery such as refrigeration equipment, done by local engineering firm; calibration of instruments, for scales and weights, done by government laboratories. The overall ranking for desired services was obtained by averaging the rankings of five managers present in the interview: Financial Manager, Plant Manager, Quality Assurance Manager, Plant Engineering Manager, and the Factory Manager. The result was: (1) energy saving, (2) productivity improvement techniques, (3) pollution control, (4) quality control testing, and (5) personnel training.

Firm 1401. Made use of a whole array of external services. For quality control: SAZ, Bulawayo Municipality Laboratories, as well as other government laboratories for microbiological tests. Also their soft-drink major international franchisor has done such testing for them. In addition, their franchisors are continuously monitoring product quality and the bottling process. They have used local training facilities for their technical and managerial personnel – as well as those provided by their main franchisor. For R&M they have used a local engineering company to repair coolers and electrical equipment. They also used calibration services for their instruments: scales, PH meters, etc. Their rankings of potentially beneficial support services were as follows: (1) personnel training, (2) R&M, (3) productivity improvement techniques, (4) energy saving, and (5) industrial safety.

Firm 1661. Have used R&M local facilities when their own shop was busy or specialized services were required, e.g. electrical work. Claimed to have used Bulawayo Polytechnic for training (interviewers assumed that they may have sponsored a student). The rankings of potentially beneficial support services were the result of averaging the rankings given by the Plant Manager and the owner: (1) productivity improvement techniques, (2) process improvement, (3) energy saving, (4) manufacture of tools, dies or fixtures, and (5) R&M.

Firm 1731. Have only used external R&M services to repair machines. The respondent only ranked three support services: (1) QC testing, (2) productivity improvement techniques, and (3) pollution control (for the smoking of meats).

Firm 71. Have utilized support service for QC, R&M, and instruments calibration. The first consisted of sample product testing, done by a monthly panel of housewives; the R&M contracted out involved fixing electrical equipment that they cannot handle themselves; instrument calibration was done every three months by local providers. Their rankings of desirable technical support services were: (1) transfer of new process technology, (2) personnel training, (3) QC testing and (4) trouble-shooting technical assistance.

Firm 1221. Claimed to have used support services for the following technical activities: QC testing, R&M, personnel training, technical assistance and instruments calibration. The QC service was provided by the Ministry of Health on a continuous, and for pay, basis. The R&M contracted out to a local supplier involved help for the yearly overhaul of main pieces of equipment. Foreign governments, through their bilateral aid programmes have helped in personnel training and the provision of other technical assistance. As to technical support services that could be beneficial for them, they singled out pilot plant experimentation and felt that it could be useful to set up a regional (for Southern Africa) experimental plant as it would not be viable to do so for a single country.

Firm 1351. Have received external technical support for R&M of oven, provided under warranty agreement by supplier. Would be interested in the following support services, as ranked: (1) productivity improvement techniques, (2) energy saving, (3) product redesign and (4) R&M.

Firm 1951. Received various support services from their shareholders, free of charge (they benefit via dividends). Such services involved: QC, R&M, technical assistance and personnel training. They paid the air-fare and per-diem for the technical personnel provided to assist them, while the shareholders paid for everything else. They would be interested in receiving the following support services: (1) product and process improvement, (2) trouble-shooting assistance, (3) QC testing, (4) productivity improvement techniques and (5) personnel training.

On the basis of the rankings shown in Table A-7, a cardinal numerical index for the support services desired was constructed, assigning the following numerical values to the ranks, 1 = 5 points, 2 = 4 points, 3 = 3 points, 4 = 2 points and 5 = 1 point. Table A-8 summarizes the results for the food industry plants interviewed and for the top nine technical support services deemed of potential benefit to those firms.

As can be seen, the areas in which the industry respondents expressed highest interest to receive support services were: productivity improvement techniques (10 out of 13 respondent firms), personnel training (seven firms), energy saving and quality control (six firms) and process improvement (five firms). The highest rankings were assigned to productivity improvement

Table A-7 Food Industry. Ranking of desired support services

Firm	*Ranking of support services*				
	1	*2*	*3*	*4*	*5*
11	Training	Lab. rel.			
211	Q.C.	Trou. sho.	Proc. imp.	Prod. imp.	R&M
241	Prod. imp.	Training	Lab. rel.	Proc. imp.	Ener. sav.
361	Prod. imp.	Proc. imp.	Ener. sav.	Pollution	Design
551	Training	Prod. imp.	Q.C.	Safety	Lab. rel.
591	Ener. sav.	Prod. imp.	Pollution	Q.C.	Training
1401	Training	R&M	Prod. imp.	Ener. sav.	Safety
1661	Prod. imp.	Proc. imp.	Ener. sav.	Tool man.	R&M
1731	Q.C.	Prod. imp.	Pollution		
71	Tech. tran	Training	Q.C.	Trou. sho.	
1221	Pilot pl.				
1351	Prod. imp.	Ener. sav.	Prod. des.	R&M	
1951	Prod. proc	Trou. sho.	Q.C.	Prod. imp.	Training

techniques, followed by training and process improvement, and then quality control.

5. Training

This section of the questionnaire distinguishes between operative workers, technical and middle-level personnel, and professional personnel. For operative workers the training needs were divided assuming that three different types of workers could potentially have been hired: (i) formally trained in the particular job (e.g. by an apprenticeship programme), (ii) workers with experience

Table A-8 Food Industry. Overall ranking of support services

Service	*No. of firms ranking it*	*Average score*
Productivity improvement	10	3.9
Training	7	3.6
Quality control	6	3.5
Energy saving	6	3.0
Process improvement	5	3.6
Repair and maintenance	4	2.0
Labour relations	3	2.67
Pollution control	3	2.67
Trouble-shooting	3	3.34
Average	**5.4**	**3.33**

acquired in similar jobs, and (iii) other workers, i.e. those without formal training and/or experience.

It turned out that firms differed substantially in their hiring policies. Some were very reluctant to hire 'educated' or experienced workers and preferred to provide their own on-the-job training. With respect to higher-level technical personnel, since many firms do not employ such personnel at all, their training needs were not perceived to be a problem.

Firm 11. For trained and or experienced workers, they estimated that on-the-job training would last between six and 12 months. For other workers, a two-year duration training period was envisaged. They did not provide training for middle-level technical personnel, and no professionals were employed by the firm at interview time.

Firm 211. This firm professed to hire their workers via the Ministry of Labour and to demand a minimum of five 'O' level examinations. The brightest were steered to middle management careers within the firm. Their on-the-job training was estimated to last up to three months; this was also the duration of their probation period. They had a more or less prescribed path for advancement within the plant which included working in the following tasks of increasing responsibility: materials handling (meat), packaging, preparing orders, mincing, weighing, cutting (band sawer), slicing, and order-calling work at a scale. No professionals were employed at the time of the interview. They have provided locally imparted management training courses to their middle-level personnel.

Firm 241. The owner and Managing Director held the view that no education was required to work in his factory. Most new employees are relatives of present workers. All start as packers or similar low grade jobs. Little on-the-job training is required according to the respondent. No training is provided for professional personnel (there is only one, the accountant). They do sponsor middle-level technical personnel to study at the Harare Polytechnic in order to upgrade their technical skills and to obtain certificate or diploma credentials.

Firm 361. Newly hired workers come from the street, but must have approved a minimum of three subjects at 'O' level. They provide a one-day induction training session, and then in-house training with courses that vary in duration according to the job. Their policy foresees the possibility of sponsoring external training for professionals, but its application will depend on the relevance of the programme of studies to the firm's needs. Middle-level technical personnel had been, and were being, sponsored to carry out technical studies abroad. A food technologist went recently to Natal for four months, and another technician is going this year, also to South Africa, for a two years' stay, to study for a diploma in food-science.

Firm 551. New workers were generally hired on the recommendation of present plant workers. Minimum educational requirements were four–five 'O' level passes. On-the-job training lasts for about six months and everybody starts with packaging and labelling spices. No professional or technical personnel were employed at interview time.

Firm 591. Incoming workers should have approved a minimum of four subjects at the 'O' level. 'Casuals' or 'on Contract' workers were generally people that had been previously employed by the firm, thus their training needs were considerably reduced. Depending on the job and prior experience, on-the-job training would last between one and six months. They were sponsoring the MBA studies of the Financial Manager at the University of Zimbabwe. Classes took place during the evenings and blocks of release time were granted as needed. They also provided training for their middle-level technical personnel to upgrade their managerial and technical skills. These courses were provided in-house (i.e. headquarters' training), and in local institutions such as the Harare Polytechnic.

Firm 1401. This firm is quite concerned with the labour training problem, to the point of operating an in-house adult literacy programme. They preferred to hire people already familiar with their machines. Used past records to track down employees that had previously worked for them under contract. Also maintained contact with Labour Department for potential candidates. For higher skill occupations they advertised in the local media. This applied to management and workshop vacancies. They provided training to their professional and technical personnel. They would pay for completion of studies in the latter case, and also send their employees to courses on management, computers, etc., imparted in local institutions.

Firm 1661. They did not hire educated or experienced workers. Tried to hire family members of present workers; also employed handicapped persons. Workers started in cleaning, later progressing to other departments. On-the-job training lasted for six months at least. They did not provide training to their professional or middle-level technical personnel.

Firm 1731. The only skilled jobs were to operate the machines. The respondent was not aware of the on-the-job training requirements. They did not provide training for middle-level technical personnel, and did not employ any professionals.

Firm 71. On-the-job training for formally trained or experienced workers lasted for one week. For other workers, three months were required. They provided training opportunities for both professional, and technical and middle-level personnel. They had recently sent one professional for a three months' course in marketing, and sponsored the apprenticeship programme of a technician. This firm had an old labour force and there was conflict between old and newer, better educated, workers. They felt that to adapt to the new economic environment they would need to introduce changes that require improving the skills of their labour force.

Firm 1221. They provided training for all personnel levels. For their professional, technical, and middle-level personnel they offered training opportunities locally and abroad (Europe), mainly to provide them with 'operative experience'.

Firm 1351. There was no on-the-job training for experienced or formally educated workers. For other workers, three–four weeks were required. Neither professionals nor technicians were employed by this firm.

Firm 1951. Their policy had been to employ highly capable, albeit untrained, personnel. They carried on a lot of in-house training and had sent technical personnel for training abroad. For formally trained, or experienced, workers, no on-the-job training was provided. For other workers, it took a year of training. Preferred to hire workers that could rise through the ranks. As an example, a young worker who had started out carrying boxes became a computer operator in a year's time. They had sent professional and technical personnel to South Africa to learn wine production and quality control methods.

E. Technological development plans

Firms differed in their attitude *vis-à-vis* the future of the industry and their place in it. Some had definite investment plans for expansion and or changes in technology; others took a wait-and-see attitude. Only responses indicating the existence of concrete plans for the future were taken into account.

Firm 11. Their plans called for diversification by increasing biscuit production and industrial catering. Introduction of a new line of frozen foods was also under study. They were considering upgrading the technology in bread-making and confectionery by leasing second-hand semi-automatic equipment. This might result in a doubling of output. Their plans foresaw the need to: (i) acquire the equipment mentioned above, (ii) hire a well-qualified maintenance engineer, and (iii) obtain a manufacturing licence for the frozen foods. Their motivation for these changes was to increase output and market power. They did not plan to hire any other local services. In addition to adding new products (frozen foods), they expected that the above changes would also improve the quality of existing products, introduce new production processes, and lead to higher productivity and lower costs. They did not foresee the need for long training or learning periods when introducing the new technologies, and did not expect to require external assistance. No organizational changes were envisaged as a result of the technology changes. Detailed cost figures were not available, but they were definitely inclined towards a lease instead of purchase of the equipment. They expected their quality to go up, and the cost to go down, as a result of the planned changes. Have had meetings with other producers to 'allocate markets'. They were looking to increase their share of the urban market, and perhaps to monopolize it, by pushing the other large bread manufacturer to supply the rural areas.

Firm 211. Firm planned to carry out minor changes, including the addition of three rooms for the smoking of meats, fish, poultry and beef. Also planned to change one machine (bowl-cutter) which would result in better capacity utilization. No local technical services were to be hired. There would be need to train the personnel involved in smoking operations. New recipes and mixtures might also be added with technical assistance from South African supplier. The smoking installation was estimated to cost less than Z$ 5,000 (about US$ 770).

Firm 241. This firm was planning to expand and to invest in technological upgrading. Purchase of a new biscuit manufacturing oven-line (the third) was under consideration, and they were also looking into the acquisition of a new line for the production of chewing-gum. The motivation for these changes was to expand production and introduce new products. No local technical services would be hired. They expected to reduce their costs and to increase the quality of their products as a result of these changes. Suppliers of the new equipment were expected to provide assistance for the installation and trouble-shooting of the machinery. The chewing-gum raw-material supplier was ready to provide the technology 'free of charge'; they also claimed to have their own formula. Local personnel would be trained by the supplier's technicians. They had estimated the cost of the chewing gum line in Z$ 1.5 million, (approximately US$ 230,000).

Firm 361. They were planning to consolidate the business and to introduce improvements as they replaced outdated equipment. They had a continuous programme of capital expenditures plus R&D. More specific information was not available.

Firm 551. Information on future plans was not available to the respondent.

Firm 591. This MNC subsidiary was planning to expand and to invest in technological upgrading. Although several technological improvements were being planned, the most important would be the introduction of a new plant and process for dried-powder instant milk. The new process would be transferred from Switzerland and their present building would be modified to better fit the new equipment. Local services would be used for the installation. While the MNC had experience with this process, local personnel would receive on-the-job training. They planned to develop their own network of dairy farmers to supply them with fresh milk. They were probably concerned about the monopoly power that the Dairy Marketing Board (their competitor) could exert. The total cost of this change was estimated at about Z$ 1.5 million (US$ 230,000 approximately).

Firm 1401. No firm plans due to prevailing economic situation. Bulawayo, where they were located, was strongly affected by the 1991–92 drought.

Firm 1661. Although the Plant Manager would like to get new machinery, and planned to go to Europe to get new ideas, there were no firm plans at the moment, and he characterized the situation as fighting a battle to hold their ground.

Firm 1731. They were planning a new factory in another location. The present one would concentrate in bottling of water, and the new one in canning of fruits and vegetables. Planned to lease the new building and estimated the cost of new machinery at about Z$ 200,000 (some US$ 31,000). The work-force required was estimated at about 80. The canning products were expected to be exported, and they were getting help from ZIMTRADE to obtain orders from countries in the Far East.

Firm 71. This firm had successfully gone public recently. Like many other local firms, having lacked access to foreign exchange for a long time, they learned to improvise which worked well in the short run and in the previous economic environment. Prior emphasis had been on increasing sales although the plant gradually become decapitalized. The opening of the economy represented a new challenge for them and they were responding by planning to expand and to invest in technological upgrading. They were thinking of totally refurbishing the plant. Their plans involved the acquisition of new equipment and the signing of a technical assistance agreement. They expected that the technical changes to be introduced would result in better quality for existing products, higher productivity, and cost reductions. They expected to provide in-house training to plant personnel on how to operate the new machinery. Expected to receive technical assistance from Danish private firm. Hopped that once the technical changes had been carried out they would be able to meet EEC quality standards. Felt that their costs were approximately 40% lower than those of international producers. Quality-wise their main need was seen in improving packaging.

Firm 1221. At difference with its private sector competitors, this para-statal firm had continuous access to foreign exchange. They had also received a lot of free foreign technical assistance. Alleged to have no plans for technological development due to lower milk production.

Firm 1351. Firm planned no changes in output capacity or technology.

Firm 1951. Although they recognized the need to improve efficiency as well as the packaging and presentation of their products, no changes were planned in capacity or technology at interview time. The change in economic environment had affected their previously paternalistic labour policies. These had been predicated on the existence of very high profits during the closed economy period. Labour force size had decreased, by attrition, by 40% approximately. In spirits (liquors) they were concentrating on the products with large production runs and discontinuing the others. For some of the latter they had become importers instead of producers. They were not happy with the wines line in which they expected to be able to compete with South Africa only at the lower end of the market. Although before they had plans to change the bottling line, and to improve quality, while also speeding up the production of wine, they had now scrapped the plans about bottling, and had slowed down their wines development plans. Would now be looking mainly for help in marketing.

F. Technical skills

Attempts to classify operative workers into skill groups raises a number of problems. A large measure of concordance seems to exist at both extremes, the high and low skills occupations, as shown for example in the ratings for tool-and-die-makers, mechanics, and electricians, at the upper end, and for movers, janitors, cleaning personnel, and helping hands at the lower end. On the other hand, substantial disagreement exist about many of the occupations falling in between (which, of course, also vary significantly in their nature according to industry). A job considered to be skilled by one respondent is often deemed unskilled by another.

Thus, in this section we concentrate only on professional personnel, i.e., staff holding a university degree, and middle-level technical personnel defined as those holding a certificate or diploma. Corroboration of the skill level is less controversial in these two cases, and information about these types of labour is important to characterize the technological capability of respondent firms. Nevertheless, it should be noted that a variety of post-secondary vocational courses with different duration and scholastic requirements exists, demanding that the information about certificate and diploma holders be handled with care on this account.

Firm 11. There were no professionals. Two vehicle-repair mechanics with certificate, one expatriate food technologist with diploma, and one sales and management expatriate with certificate. Total number of employees, including management was 126.

Firm 211. No professionals were employed. One diesel motors mechanic with certificate, one food technologist with diploma from the Meat Institute, two certificate holders in personnel and training, and two diploma holders in accounting. Total personnel, including management: 54.

Firm 241. One accountant with MBA degree who was trained abroad. One machine-fitter with diploma and one electrician with certificate. Total number of employees was 364.

361. This firm posed a problem in determining the number of highly skilled personnel, because some managerial functions were carried out by the Group to which the plant belongs, and at the same time some high-level plant technical personnel were also Directors for the whole Group. It was attempted to consider only personnel that was clearly assigned only to the plant visited. There were no engineers among the professional personnel, but three chemists (one trained locally and two abroad), two accountants, locally trained, one information systems person, also trained locally, seven food-science staff (four trained locally and three abroad). Among the middle-level technical personnel: two food technologists with diploma and one with certificate (one of them expatriate), two mechanics, one with certificate and one with diploma, and 10 R&M artisans with certificate. The approximate total number of employees was estimated by the respondent to be 800.

Firm 551. One economist, with MA, trained abroad (Managing Director and owner), one bookkeeper and one sales person with certificates. Total number of employees was 49.

Firm 591. Information about high-level skills was supposed to be provided by Financial Manager after the interview and delivered to the university. Total number of employees estimated at 150.

Firm 1401. One engineer (trained abroad), one accountant and one MBA, one marketing staff (trained abroad) and two personnel staff, one trained locally and one abroad. At the middle level: two diploma holders in food technology, one diploma holder in electrical work, one diploma holder in data processing. Total number of employees was 360.

1661. One certified accountant. Among middle-level technical personnel: three repair and maintenance journeymen with certificate. Total employment was 312.

Firm 1731. One diploma in accountancy, and one certificate in secretarial work. Total personnel 21.

Firm 71. One engineer, two chemists, four accountants (one trained abroad), one social science major, two business administration majors, three sociology majors, and one graduate in animal science, among professional workers. Among middle-level personnel: one with diploma in wood-working, two diploma holders and four certificate holders in food technology, three electricians with certificate, one diploma holder and three certificate holders in data processing, and three diploma holders in marketing. Total personnel estimated at 1131.

Firm 1221. One engineer, two chemists, two accountants, and four with other majors, all locally trained. Among mid-level technical personnel: one diploma holder and 29 certificate holders in metal-working, five diploma holders and four certificate holders in food technology, 15 with certificate in electrical work, 199 (sic) with certificate in data processing, and three in other fields. Total number of employees: 839 (including 19 apprentices).

Firm 1351. Owner has a degree in teaching. No other professional or middle-level technical personnel. Total personnel: four including the owner.

Firm 1951. Two chemists (one trained locally and one abroad), one accountant (trained abroad), two MBAs (trained abroad), one major in human resources (trained locally). Among middle-level technical personnel: two certificate holders in wood-working, one with diploma in electrical work, one in data processing, and four mechanics with diploma. Total number of employees: 312.

Table A-9 below summarizes the information on skilled personnel.

Table A-9 Professional and middle level skills

Firm	*Profess.*	*Diploma*	*Certifi.*	*Employ.*	*% Prof.*	*% D + C*
11	0	1	3	126	0.0	3.2
211	0	3	3	54	0.0	11.1
241	1	1	1	364	0.002	5.5
361	13	3	12	800	1.62	1.87
551	1	0	2	49	2.04	4.08
591	n.a.	n.a.	n.a.	150	n.a.	n.a.
1401	4	3	0	360	1.11	0.83
1661	1	0	3	312	0.32	0.96
1731	0	1	1	21	0.0	9.52
71	14	7	10	1131	1.24	1.50
1221	9	6	250*	839	1.07	1.50
1351	1	0	0	4	25.0	0.0
1951	6	6	2	312	1.92	2.56
Total	**50**	**31**	**287***	**4,522**	**1.10**	**7.03***

Notes:
n.a.: Information not available.
D: Diploma
C: Certificate
* Information likely to be distorted by high number of personnel with data processing training (probably quite limited in nature), reported by firm.

Appendix C: Firm Interviews by Sector: Textiles and Clothing (ISIC 32)

Eight out of the 10 firms surveyed manufacture clothing (children's, women's, menswear) and uniforms while two combine garment-making with textile activities – one is involved in printing cloth and another manufactures cleaning rags and also engages in knitting.

A. Background information

1. General

Firms in the sample vary significantly in terms of size (measured by employment), output per worker, ownership, and ethnic origin (see Table AII-1). The range of output per worker varies between $ 1,600 and $ 19,700. – Reflecting differences in capital intensity, i.e. the fact that more equipment per worker is required to manufacture textiles than clothing production, the two plants engaged in textile processing and rags manufacturing have the highest output per worker values, while correspondingly, firms only manufacturing garments tend to have lower such values.

The broad headings textiles and clothing encompass a great diversity of manufacturing activities. The clothing firms produce a variety of products (trousers, shirts, jackets, suits, dresses, skirts, etc.) both for the domestic market and for exports. Due to the impact of competition after the economy's opening, some have shifted to the production of more specialized clothing such as school and other uniforms. One of the clothing firms does subcontracting work while a large majority manufactures on the basis of their own or acquired designs.

Table AII-1 Textiles and Clothing. Basic characteristics of firms interviewed

Firm	*Product*	*Employ.*	*Ownership/ethnic*	*Location*	*Sales/worker (US$)*
51	Menswear	450	White	Harare	7,000
681	Ladieswear	230	n.a.	Harare	8,700
751	Subcont.	11	Asian	Bulawayo	2,100
1701	Clothing	95	African	Harare	1,600
161	Childwear	3	African	Harare	3,300
271	Print/clo.	1554	White	Harare	11,900
461	Clothing	13	African	Harare	n.a.
1121	Rags/knit.	47	White	Bulawayo	19,700
1581	Lad./ch.cl	156	White	Harare	2,500
1681	Clothing	49	White	Harare	6,800

2. Market orientation

In this industry one would expect Zimbabwe to be competitive in the international market, particularly in the clothing segment, which depends heavily on low-wage, semi-skilled labour. Table AII-2, showing exports as a proportion of total sales, indicates that four of the 10 firms do not export at all. Three of the non-exporting firms are also the smallest in terms of value of sales. Three firms export 20% or more of their output and one of them, the largest, exports about two-thirds of total output. The largest firms also have the largest shares of exports in total sales. Several respondents indicated that local sales were made at prices substantially higher than international prices, i.e. there was room for price discrimination.

3. Organization

The questionnaire inquired about the existence of professional management, the type of organizational structure adopted, and the presence of specific, formally assigned, managerial functions. As can be seen from Table AII-3, only half of the firms surveyed had professional management. Of these, four had one or more tiers of managerial functions assigned. The larger firms tended to have professional management and one or more levels of functional management.

B. Entrepreneurial history and acquisition of technology

1. Origin of the firm and initial technology

Firm 51. Old clothing manufacturing family business with a majority of the shareholders residing abroad. Originated as wholesale business in 1940 and in 1960 broke away from wholesale. At that time it already had about three-quarters of present machinery and was about two-thirds of its present size. Moved to a bigger building in 1976. Their products were meant for the middle-income market. Profess to manufacture good garments (menswear) at reasonable prices. Exports to Europe begun as early as 1960. Following UDI they started exports to South Africa.

Table AII-2 Textiles and Clothing. Proportion of output exported

Firm	*Exports ($US)*	*Sales ($US)*	*% Exported*
51	524,000	3,169,000	16.5
681	461,000	2,000,000	23.0
751	0.0	23,100	0.0
1701	0.0	148,000	0.0
161	0.0	9,800	0.0
271	12,136,000	18,500,000	65.6
461	0.0	n.a.	0.0
1121	50,000	924,000	5.4
1581	800,000	3,846,000	21.0
1681	212,000	2,160,000	9.8

Table AII-3 Textiles and Clothing. Organizational structure

Firm	*Profess. manag.*		*Organizational structure*				
	Yes	*No*	*Direct (1)*	*Super. (2)*	*1 Tier (3)*	*Tiers (4)*	*Other*
51	X				X		
681	X					X	
751		X	X				
1701		X					(2),(3)
161		X	X				
271	X					X	
461		X	X				
1121	X			X			
1581	X				X		
1681		X					(1),(3)

Since 1983 they have cut exports to South Africa, and in recent years have also reduced their exports to the UK and Italy. Products exported were trousers and shirts (made of cotton). They claimed that although Far East producers enjoy production subsidies, they could compete. Because they do not pay duties in Europe, they have a 13% advantage over non-African exporters. They do not enjoy protection in the domestic market, but claim that with a 40% tariff on imported fabrics the textile producers would be well protected. They would also like to start with a 35–40% tariff that could be removed gradually. In exports they just break even, and claimed to have turned down export orders. They would use the time afforded by protection to re-equip. Need to buy some $350,000 in machinery. They also perceived the need to train technical management and mechanics. So far they have only had on-the-job training. They see the formal training as required to compete in international markets. Recent acquisitions of equipment were specified by their technical manager. Started with a small machine making uniforms for wholesale and developed gradually. No additional information was available about the initial equipment. It was assumed that there was no formal pre-investment or project study before starting the firm.

Firm 681. Firm was established 34 years before by a European entrepreneur who had other factories manufacturing ladies wear in South Africa. Facing serious difficulties in 1983 and about to be liquidated, the government stepped in and through the Industrial Development Corporation (IDC) acquired 51% of the stock. The rest of the shares were held by the original owner and his partners. Their exports are mainly suits, jackets, ladies' and boys' trousers. Have their own designers, and keep abreast of fashion trends, new materials, etc. Firm employs an experienced, British born and educated, deputy general manager in charge of merchandising, who negotiates herself most of the export orders to the UK. Their exports benefit from draw-back allowance, but they complained about not being paid the 9% subsidy to which they were entitled. Admitted that there was substantial price discrimination between exports and domestic sales. For example, a pair of shorts was sold locally at Z 56, but only fetched Z 32 in the export market; a suit was sold for Z 155 locally and for

Z 88 as an export item. Both examples involve a 43% price differential, approximately. The firm started small, with equipment selected by the original entrepreneur, and no pre-investment study seems to have been carried out. At the time of change in ownership, the IDC carried out a project appraisal using their own staff for that purpose. More recent acquisition of equipment (30 Japanese sewing-machines worth approximately US$ 62,000) was specified by the firm on the basis of prior experience, and having seen similar machines in operation elsewhere.

Firm 751. Born in China from Chinese father and Indonesian mother, female entrepreneur now operating small cutting, making and trimming (CMT) operation which she supervised directly. Back in 1964 she was running another company owned by Europeans who left the country after UDI. Enrolled part-time in dress-making lessons at the Polytechnic, but was also trained as a nurse. Took the UK equivalency examination and worked at Memorial Hospital before starting her own business. Back in 1985 she was able to accept orders to produce her own line of clothes. Presently only did subcontracting work because she could not afford to buy her own materials; categorizes the market as unpredictable. Also did some private dress-making but stopped because it was not lucrative enough. Started with four used machines acquired from companies that were closing in the early 1980s. There was no formal study before initiating business. All her equipment is barely worth around US$ 10,000.

Firm 1701. Company was set-up to make overalls and other heavy-duty uniforms for the government, and at times also for other buyers. All sales were domestic. After restructuring of the government purchasing authority they lost their special status and now had to compete in open tenders. Government was still the main customer. In trying to diversify they are also doing CMT. Government also supplied materials for its orders. Had to prepare study to get bank loans. It was done by their own accountant. Acquired a going concern with all its machinery, building, etc. There are four partners, one acts as general manager and the other three sit on the company's board. Present machines were suited for manufacture of police and other uniforms, made of coarse and tough cloth. They had difficulty adapting this equipment to handle other type of fabrics, as required for school uniforms, for example.

Firm 161. Seamstress owner operates shop with only two workers. Started in 1976 working at home. At present location since 1985. They produced school uniforms, baby sets, children's dresses, and hats, all for the local market. There was no initial study and the entrepreneur bought the sewing, and latter knitting, machines, based on her experience. They had to reduce the scope of their operations because of building use restrictions. Had other machines stored away. Complained about lack of financing.

Firm 271. Started as small family business, some 35 years ago, with only a few sewing machines. Presently were employing more than 1500 people in the manufacture of printed cloth, mens and ladies garments. A substantial proportion of the garments was produced for export. There was no initial pre-investment study and the entrepreneur selected the equipment. More recent machinery acquisitions were fully specified by the firm on the basis of information gathered in trips abroad, catalogues, and the attendance of trade fairs. Machinery was fairly new. At interview time they were experiencing difficulties due to macroeconomic conditions. Produced for export to the US garments sold under well-known foreign labels. Importers oversaw the plant to ensure that their quality standards and other specifications were met.

Firm 461. Started in 1978 after having been employed as a machinist in another firm. In 1984 he had four employees, and now had 12 at the plant and four at sales outlets he opened because the retail stores were taking too much time to pay him. Learned from experience and six months of training in college. There was no initial pre-investment study. All machines were selected by the entrepreneur and bought used from large factories. Was producing school uniforms and protective clothing, all for the domestic market.

Firm 1121. Plant established in 1951. The machinery for waste-processing was more than 40 years old and it was bought used from the UK. Added knitting operations more recently and their machinery is newer (some 15 years old). Before acquiring the knitting machinery they visited another factory to see it in operation. The company started as a partnership to recycle cotton waste to make cleaning rags and upholstery fillings. The original partnership was dissolved and the firm sold to a local group. A small proportion of the cleaning threads and rags was exported to Botswana where no duties were imposed. Complained about South Africa's punitive import tariff.

Firm 1581. Firm was started by father of present general manager in 1968, with six machines bought after the closing of another firm. He had been an agent of a clothing firm and decided to start his own business. Came to help her father in 1969 and began making women's dresses. During the 1982 recession sought help from German technical assistance, and started to produce children's clothing for export. Exports were made at marginal cost, but they also enjoyed a 9% export subsidy which they complained was generally paid with long delays. Price differentials between exports and the domestic market exist. For example, for women's dresses while the export price was Z$ 33, the local price was Z$ 55, i.e. about 40% higher. Initial equipment was selected by entrepreneur and there was no pre-investment study.

Firm 1681. Family business that produced mostly ladies' underwear. Father, mother and son managed the operation together. Started back in 1980 with four second-hand machines and capitalized the firm gradually acquiring equipment with good-terms loans. Entrepreneur (father) bought the equipment without having carried out any pre-investment studies. More recently, machines had been acquired interfacing with suppliers, and also by examining catalogues and what was available in the market. In the past, the government allocation of import licences was inefficient and the system was quite corrupt. They exported about 10% of total sales at prices slightly below those in their domestic sales.

2. Source and characteristics of equipment

Since it was not possible to survey every piece of equipment in each of the plants visited, respondents were asked questions about their three principal machines.

As can be seen from Table AII-4, a large proportion of the machines originated in Japan, with Germany and the USA among other major sources. The UK and South Africa were also mentioned as sources in a few cases. The equipment was originally acquired in about equal proportions new and used. Although some cases of recently (one to six years old) acquired machinery were also

Table AII-4 Textiles and Clothing. Information about key pieces of equipment

Firm	*Origin*	*Age (years)*	*Bought new or used?*	
			New	*Used*
51	USA	16	X	
	Japan	>30	X	
	Japan	4		X
681	Japan	22		X
	Japan	1 and 22	X	X
	S. Africa	15 and 22	X	X
751	USA	29	X	
	Japan	3	X	X
	Japan	10		
1701	Japan	n.a.		X
	USA	n.a.		X
161	USA	18	X	
	Japan	12	X	
271	Japan	15	X	
	Japan	av. 10	X	
	Japan	av. 10	X	
461	USA	>40		X
	Japan	>40		X
1121	UK	40		X
	UK	15	X	
	Germany	15		X
1581	Japan	25	X	
	Germany	4	X	
1681	Japan	2	X	
	USA	10	X	
	Germany	6	X	

reported, a majority of the equipment was 10 or more years old, and in some cases 30 years old.

3. Use of foreign technology

Here we asked whether the enterprise had a foreign licence or technical assistance agreement, if any help was received for commissioning the plant or installing the equipment, and, if such had been the case, what was involved in the assistance provided.

As shown in Table AII-5, three of the 10 firms had licences, and/or technical assistance agreements, while only one firm required outside help to instal the equipment. This may be explained by the fact that a large proportion of the

Table AII-5 Textiles and Clothing. External technical assistance

Firm	Licence		Technical assist.		Outside help	
	Yes	No	Yes	No	Yes	No
51		X		X		X
681	X		X			X
751		X		X		X
1701		X		X		X
161		X		X		X
271	X		X		X	
461		X		X		X
1121		X		X		X
1581		X	X			X
1681		X		X		X

firms were dedicated to clothing manufacture requiring generally light and easy to instal machinery. In fact, many were just sewing-machines and variations thereof with little in terms of controls and thus complexity.

Firm 681. This firm had a technical assistance agreement, drafted in 1982 and implemented one year later, with a minority shareholder, residing in South Africa, who was the previous owner. The agreement allowed them to get patterns and samples, but respondent did not consider it to have been very helpful. They used to receive regular plant visits in the past, and they also travelled to South Africa to avail themselves of information on the range of products available and the fabrics in use. They paid a sliding royalty on sales as follows: 3% on the first Z$ 2.5 million of sales, 2% on the following million, and 1% thereafter. The agreement precludes exports to South Africa which were difficult in any case due to their high import duty. They had recently installed without difficulty and no outside help some 30 machines imported from Japan.

Firm 271. This firm produced garments under well-known foreign brand labels. The importer oversaw the plant to make sure that the output met their standards. There was no charge for this service. They received help from the supplier to instal new equipment.

Firm 1581. Received extensive technical assistance under Germany's bilateral foreign aid programme. Used to get it free of charge; recently they have had to pay part of the costs involved in seminars, trips of experts, and other related expenses. No help was required to instal equipment.

C. Operating capabilities

1. General

Firm 51. Premises well adapted to industrial production; well-lighted and large enough plant, but flow of production did not seem to be well organized. In contrast to a majority of textile and clothing plants, most of the workers were male with very few females on sight. The equipment appeared to be very old, and the plant was working one shift at about 75% capacity utilization. Machines could be idle depending on the product(s) being manufactured. Unused capacity also aroused due to lack of certain imported materials.At visit time they were missing imported shoulder pads. They had not arrived due to an administrative error which placed these items in the wrong import list.

Firm 681. Very well-lighted plant with clean floors, but solvents and paint were being used without proper ventilation or air-extraction system. Most of the equipment was over 20 years old. Plant worked one shift at about 80% capacity utilization. Some idle workers. Some idle machines either waiting for repair or to be installed. Also they were not using some machines capable of doing fancy stitching due to lack of orders for such work.

Firm 751. Crowded space with poor lighting. Machines were quite old (one goes back to 1902) and some were not being used due to changes in demand and thus output mix. Worked one 45 hours/week shift.

Firm 1701. Appropriate plant building, clean and well-lighted, but with a lot of unused space and equipment. Equipment is generally old and some machines were out of commission to be fixed or serviced. Due to falling demand some personnel had been recently laid off and plant was working one shift at about 50% capacity utilization.

Firm 161. Clean, small shop. One room operation in office building that operated in one shift at about 90% capacity utilization.

Firm 271. Modern, clean, well-lighted and organized plant. Equipment was quite modern with less than 15 years of age generally. Capacity utilization at time of visit was about 60%, but it fluctuated greatly with demand which had declined recently leading to downsizing in some departments. Some orders, for export for example, were seasonal. The plant worked one shift with occasional overtime, but some sections could work two and three shifts if required.

Firm 461. Poorly kept plant. Very untidy, with little or no space to walk. Shop was in a small basement room with good windows and a fan for air extraction. Very old equipment (more than 40 years old). Worked one shift at about 80% capacity utilization.

Firm 1121. Spacious and well-lighted new premises. Some machines had not yet been installed and the storage of inventory seemed to lack organization. Whole plant worked one shift. The knitting sections generally worked three shifts, but they were lacking in sufficient demand. Capacity utilization was below 100% during the year.

Firm 1581. Well-organized, clean, well-lighted plant with good working conditions. Operated one shift at about 70% capacity utilization.

Firm 1681. Very clean plant in new building with good, flexible, layout. Machinery was quite new, ranging in age from one month to 13 years. Plant worked one shift at almost full capacity.

2. Troubleshooting

Firm 51. Have experienced equipment breakdowns and quality problems. Machines were fixed with local help. They saw the need to study whether to further repair or to replace existing machinery. They experienced periodic quality difficulties. Most recent problem was with shirt collar pressing. After customer complaints worker was retrained.

Firm 681. Suffered frequent steam-boiler breakdowns and had difficulty obtaining spare parts. Repairs were done in-house. Also had difficulties with button-hole machine now replaced.

Firm 751. Machines broke down occasionally and were fixed by local firm when in-house personnel could not solve the problem.

Firm 1701. Experienced machine breakdowns. They were fixed by in-house mechanic with outside help when necessary.

Firm 161. No problems were reported.

Firm 271. Equipment breakdowns were not serious and were fixed by in-house personnel.

Firm 461. Their main problem was equipment breakdowns, mostly due to the old age of many machines. What could not be fixed in-house was done in local specialized shops, or by local technicians in the plant in the case of heavy equipment. The breakdowns of the old equipment also gave rise to quality problems and resulted in bottlenecks or imbalances in the production process. Their hemming and button-hole and button sewing machines were not used all the time because they produced more than the regular sewing machines. However, they were not very concerned about idle machines since due to the relatively low cost of the equipment their down-time did not entail much higher costs.

Firm 1121. Have experienced safety problems. One machine had a high propensity to catch fire. They have installed dust extractors and trained personnel how to react to the problem.

Firm 1581. Only experienced minor breakdown problems.

Firm 1681. Equipment breakdowns were handled by the owner and his personnel with occasional resort to outside local help. Bottlenecks in production process have not been solved. New, expensive, machines would be required.

3. Quality control

Firm 51. Visual, practical inspections were applied to the final product, during the production processes, and to materials and components in process. Purchases of raw materials and components were not inspected because the 'fabric supplier is a monopoly', and they had no alternative supply source. They estimated that about 6% was lost due to defects in the material and had to be cut off. They did not follow any norms or standards. There was no sampling routine for testing during the production processes. They assumed that their quality was similar to that of other local producers, and had no information about the quality level of imports. They estimated that they had a 0.5% rate of rejections due to internal production deficiencies, without counting any fabric problems. As an example, during the last month they had returns worth Z$ 6,000 out of Z$ 3,300,000 billed, i.e. less than 0.2%. On the other hand, fabric-based rejects were about 3%. Customer complaints were handled via direct contact with them and also through the sales force. Although they did not have a statistical quality control (SQC) system, they had plans to introduce quality specifications, charts, and sampling.

Firm 681. Final product, production processes and raw materials purchases were all subjected to visual, practical inspection. Imported fabrics with defects were difficult to return but claims were made to foreign supplier. During manufacture, quality is the responsibility of the production supervisor. After packaging, the final product is inspected by quality control personnel. They used no norms or standards but the various garments were produced according to specifications. No sampling routines for inspection existed. They felt that their quality was higher than that of other local producers, but were afraid of the competition of imports from South Africa. Claimed to have had an internal rate of rejection well below 3% last year. They had seen an improvement due to a staff training programme for new employees. The average age of new hires had also been substantially reduced. Customer complaints were directly channelled from dispatch to top manager. Products in question were replaced or credit was granted if warranted. No system of SQC existed.

Firm 751. Carried out visual, practical inspection of final product, production processes and purchases of raw materials and components. Materials and components in process were not inspected. One staff performed the quality control and the manager also checked quality. No norms or standards were in use and there was no routine for sampling inspection. They assumed that their quality was equal to that of other local producers and had no information about imports. Estimated that internal rejects due to poor workmanship were below the 1% rate last year. Claims were handled by direct contact with the customers and they repaired the garment(s) involved. There was no system for SQC.

Firm 1701. Final product, production processes and raw materials and component purchases were the object of visual, practical inspection. No use of norms or standards and no sampling routine was in place. They were unaware of how their quality compared with that of other producers. Rejects due to raw material problems were quite low, about 0.1%, but those due to poor workmanship could be as high as 15%. Blamed those on the lack of well-trained workers and appropriate machines. There was no improvement trend because part of their skilled personnel had been lost due to retrenchment. Customer complaints were handled via the sales force addressing the problem first through the sales manager, and then through the factory supervisor. No SQC system was in place.

Firm 161. Only inspected, visually and practically, raw materials and components purchases. No norms or standards were in use and they did not have a routine for inspection by sampling. They assumed that their quality was lower than that of other local producers, and much lower than imports. Last year's internal reject rate was estimated at less than 1%. They kept track of complaints by direct contact with customers. Did not have a SQC system.

Firm 271. The final product was the subject of 100% visual inspection, as well as laboratory testing. Raw materials and component purchases were subjected to laboratory tests, and standards were applied. For printing and dyeing they also had laboratory tests. They applied national and international norms. The latter were required to export and were compulsory, while for local markets their application was voluntary. No sampling routine for inspection during the process had been established, but they did apply sampling for inspecting materials. They believed that their quality was much higher than that of other local producers and about the same as imports. Their exports competed with those of the 'four Asian tigers', and the quality of their products was about average in international markets. Their quality control effort had been strengthened. Rejects in their exports to the US were about 3% three years ago, but only 1% last year. Maintained direct contact with customers, and travelled twice a year to meet customers abroad. Did not have a SQC system.

Firm 461. Final product, processes, raw material purchases and materials and components in process, were all subject to visual, practical control. No norms or standards were in use and there was no established sampling routine for quality control during the production process. They considered their quality to be equal to that of other local producers and had no information about imports. When a new style was introduced the rate of internal rejects tended to be high (20%). These were reworked and sold as 'seconds'. With the passage of time, learning took place and the rate of rejects tended to decline to 2–3%. Customers' complaints were handled by direct contact with them. There was no SQC system.

Firm 1121. Quality considerations were not important for the cleaning rugs product, and they only inspected visually their final products. No norms or standards were in use and there was no routine for sampling inspection during the process. They believed that their quality was comparable to that of other producers, both local and foreign. Rejects were recycled and were less than 1% last year. They were in contact with their customers through the sales force to find out reasons for their dissatisfaction. There was no SQC system.

Firm 1581. Final product, production processes and raw material purchases were inspected by visual, practical means. No norms or standards were applied. Nevertheless, they were exporting to Germany and small amounts to other markets (Botswana, Malawi and Saudi Arabia). There was no sampling inspection routine. They considered their quality to be higher than both local and imported alternatives. Last year production-originated rejects were only 0.01%, but taking into account fabric defects they reached 3%. They attributed the high rate of fabric defects to local mills having become quite old and being under pressure to increase output. Reworked rejects were sold at a 10% price reduction. Customer complaints were handled directly with the clients and also through the sales force. There was no system of SQC.

Firm 1681. They did not control raw-material purchases, but subjected the final product, production processes, and materials and components in process, to visual, practical, inspection. Quality checking was done by production personnel. No norms or standards were in use. There was no sampling routine for inspection during the production process. Held their quality to be higher than that of local producers, and slightly lower than that of competing imports. Rejects last year were between 2 and 3%, and three years before around 3%. Claimed that the rejects were essentially due to defects in the quality of the cloth. Rejects were sold as 'seconds'. Made direct contact with customers in case of complaints. They sold garments with well-known brand labels. Dealt with large local retail chains. No SQC system was in place.

4. Repair and maintenance (R&M)

Firm 51. There was a R&M shop with three workers but no preventive maintenance routine. Machines were checked when they broke down plus there was a yearly plant overhaul. There was no established frequency for oiling and/or parts replacement. Equipment breakdowns were frequent: once–twice per month. The head-mechanic was in charge of R&M and reported directly to the plant manager. They did their own oiling, replacement of belts, and other simple servicing, relying for the rest on outside help. This was provided by a local mechanic on call who had worked previously for them and knew their machines.

Firm 681. Had in-house R&M shop and every year during Christmas vacation closed down the factory to check the machinery and overhaul it if necessary. While equipment breakdowns were deemed to occur frequently, no supporting information on their frequency was available. The factory manager was in charge of R&M, and they were able to carry out simple servicing and parts replacement by themselves. Lubrication and other routine servicing was done weekly. Outside help was used to service the boiler.

Firm 751. No R&M shop available, and no established routine for preventive maintenance. Machines were checked when wear was perceived. They also checked before closing down for Christmas vacation. Oiling was done weekly, but some machines were self-oiling only requiring a topping off. Equipment breakdowns were infrequent. Owner was in charge of R&M and was able to carry out simple servicing or parts replacement. Outside help was used to repair those machines the owner could not fix by herself.

Firm 1701. There was an in-house R&M shop with one staff assigned. While there was no pre-established preventive maintenance routine, they did check the machinery every weekend. Oiling was done once a week, but there was no established frequency for parts replacement. The mechanic responsible for R&M reported directly to the General Manager. They were able to do oiling and simple replacement of parts by themselves. Equipment breakdowns were not frequent and an outside mechanic was called in to assist in fixing the broken-down machines.

Firm 161. No R&M shop in the plant and they stopped to check the machines only when something broke down. Oiling was done during the weekend and they had no established frequency for parts replacement. Breakdowns were not frequent. Nobody had been assigned responsibility for R&M and they were totally dependent on outside help to fix broken-down machines.

Firm 271. Had in-house shop for R&M work and were proud of their preventive maintenance record. Thus machines were stopped for check-up at pre-established times, whether they exhibited problems or not. R&M was under the responsibility of the Production Manager. They had the capacity to strip most machines to replace standard parts or get spare parts made. Specialized machines (such as printing equipment) were overhauled once annually by outside experts, but all simple equipment overhaul was done in-house. Machines were self-lubricating; they only had to top off oil receptacles when they were low. Parts were replaced as needed. Breakdowns of equipment were not frequent.

Firm 461. There was no R&M shop in the plant. Every Friday they topped off the oil lubricating receptacles. Equipment breakdowns were frequent (about once per week). The owner was responsible for R&M and he carried out minor repairs. They used the services of a shop in town for other R&M tasks.

Firm 1121. There was no R&M facility in the plant. They only stopped to check the machines when something broke down. Oiling was done daily but parts were only replaced occasionally. Equipment breakdowns were infrequent. The Plant Supervisor was in charge of R&M. They did all simple maintenance tasks plus had the ability to strip most machines and to replace standard parts or get spares made for them. Outside help was required to obtain some spares, such as castings, and for electrical work.

Firm 1581. There was a R&M shop in the plant. Oiling was done weekly, but parts were only replaced upon failure. Equipment breakdowns were not frequent. A mechanic was in charge of R&M and he reported to the Production Manager. They performed all simple maintenance tasks plus had the capacity to strip most machines and to replace standard parts.

Firm 1681. The plant had a R&M shop. There was an established routine only for oiling the machines. Machines were stopped to check them only when something broke down, and parts were replaced only when needed. They resorted to outside help to service their machines as needed. Breakdowns of machines were quite frequent – about every couple of days. The Production Manager (owner's son) was in charge of R&M. They claimed to have the capacity to do simple maintenance tasks and also to strip most machines to replace some parts.

5. Industrial engineering

Firm 51. An organized effort to improve plant productivity had been launched recently with the arrival of the new Technical Manager. One technician was permanently assigned to the task. The Technical Manager claimed to carry out time-and-motion studies himself. He had applied the MTM method in the trousers section and found that employees were working at a 65% efficiency level. He would like to raise it to 85% by simplifying the manufacturing system. They had a bonus scheme but no piece-rate, which they maintained was illegal in the country. The bonus was equivalent to 25% or more of the salary, and was paid at year's end because then it was considered non-taxable. Thought that their output per worker was comparable to that of foreign plants, but were unaware about how they compared with local producers. Scheduling of production was rudimentary and based on delivery dates according to orders received. Inventory control of raw materials, work in progress, and final product, was carried in a cardex system. No computers were used for these tasks. They did not continuously assess the plant's layout and only looked at it in case of plant expansion or replacement of equipment. Last layout change was done some 10 years ago.

Firm 681. Claimed to have an organized effort to improve plant productivity. One staff was assigned to it. He held a technical certificate from Harare Polytechnic, and carried out work-study and time-and-motion studies. They used old, inherited, standards for all dress-making operations and did time-and-motion studies for verification purposes and to establish standards for new tasks. They had a point system for the production of garments and a bonus was granted based on the number of points attained. They considered their productivity to be lower than that of foreign plants, but were unaware of how they compared to local producers. They planned production three months ahead, and broke down the period into three weeks intervals. Production for exports required longer forward planning of about eight weeks. They carried manual inventory control for raw materials, work in progress, and the final product. Computers were used in accounting and for production planning. They planned to improve their accounting system and computer in the near future. Plant layouts were changed only to accommodate new machines.

Firm 751. There was no continuous, organized effort to improve productivity; they did not carry out time-and-motion studies and did not have production standards. Did not know how their output per worker compared with that of other local or foreign producers. Planning of production was based on the orders received and claimed that since they did subcontracting (CMT) work they were unable to carry out any planning. There was no inventory control system and no computers were in use. Plant layout was only changed in case of a move.

Firm 1701. Did not engage in an organized effort to continuously improve plant productivity and did not carry out time-and-motion studies. Production standards were based on past experience, and they were unaware of how their productivity compared with that of other local or international producers. There was no advance scheduling of production, but they did have inventory control of raw materials, work in progress and final product. No computers were in use for production planning or control, and they did not assess continuously the plant layout.

Firm 161. They did not engage in a continuous and organized effort to improve plant productivity. Did not carry out time-and-motion studies, did not have production standards, and did not know how their output per worker compared with that of other producers, local or foreign. Production scheduling was based on the delivery dates for the orders on hand. Only kept inventory control of final product. No computers were in use. They did not continuously examine the plant layout with a view to improving it.

Firm 271. They made a continuous and organized effort to improve productivity in the clothing manufacturing sections where labour costs were very important. Claimed that one engineer, one technician, and three workers were assigned to this task. While they did not carry out time-and-motion studies, when they engaged in long production runs a study was carried out to set production standards. Being by far the largest local plant, they considered their productivity to be higher than that of smaller local plants, and about 95% of that of European competitors. Production planning covered the year, the season, and weekly intervals. There was formal inventory control of raw materials, work in progress, and final product. Computers were in use and they claimed to have installed an on-line direct inventory control system. They had both a mini-computer and also PCs, and utilized a variety of business and accounting software. Plant layout was being changed continuously according to the size of export orders they received.

Firm 461. There was no continuous, organized effort to improve productivity. No time-and-motion studies were carried out, but they used bonuses to provide worker incentives to meet and surpass output targets. They believed they had higher productivity than other local plants. Claimed to have advance scheduling of production. Demand for certain products (school uniforms) was seasonal requiring longer planning (three months ahead). They had formal inventory control for raw materials and final product. No computers were used for production planning or control, and they did not continuously assess the plant layout with a view to improve it.

Firm 1121. There was no organized effort to continuously improve plant productivity and no time-and-motion studies were carried out. The Factory Supervisor and the General Manager set production standards based on experience. They were unaware of how their plant productivity compared with that of other local or foreign producers. Claimed to have advance scheduling of production with planning of production at least one month ahead. Had manual, cardex system for raw materials and final product inventory. Tried to keep about four months' stock of raw materials and one month of finished products. Complained about the extent of capital tied-up in stocks. No computers were in use. Plant layout was reassessed on occasion of expansions or replacement of equipment. They planned to move into new premises shortly, but did not expect to make any layout changes.

Firm 1581. Claimed to have a continuous, organized effort to improve productivity. Production Manager and two technicians were assigned, part-time, to this task. They also professed to carry out time-and-motion studies. Every product had a time standard assigned based on experience. They recognized that their productivity was lower than that of large local, or foreign, producers. They argued that this was due to existing labour laws that made it difficult to get rid of unproductive workers. Also considered that low wages demoralized the work-force and were matched by low productivity. Production planning was based on delivery dates for orders. Kept an inventory control cardex system for raw materials, work in progress and final product. No computers were in use. Layout was very flexible and was altered to accommodate changes in orders received. Last change was done during the year to accommodate new machines.

Firm 1681. There was no continuous and organized effort to improve plant productivity. While they presently did not carry out time-and-motion studies, they had knowledge and experience with MTM method and planned to introduce wage incentives. They had established output targets for each line. Considered their productivity to be higher than that of local plants, but only 40% of that in UK plants. Attributed the difference to the hot climate and the lack of skills. Had advance scheduling of production and claimed that longer-term planning was required (about 10 weeks ahead) because of the difficulties experienced receiving imports. Carried cardex inventory control of raw materials, work in progress and final product. Every week they had a report to account for all inventories. No computers were in use. They claimed to continuously assess the plant layout with a view to improving it. Had installed efficient power supply system allowing them to easily move machines around if required.

6. Industrial safety and pollution control

Firm 51. Responsibility for safety and pollution control had not been assigned to any office or person, there were no written rules either for safety or for environmental control, and no fire drills were carried out. Yet they rated their safety and pollution control records and procedures higher than those of other firms in the industry.

Firm 681. Responsibility for safety had been assigned to a committee that included plant workers. No office had been assigned responsibility for pollution control. There were no written rules for safety (but they had a poster), and no written rules for environmental control. They had fire evacuation drills, but only once in two years, and they had trained some of the personnel for that purpose. They rated their safety and pollution control records and procedures as about the same as those of other firms in the industry. However, during the visit it was observed that solvents were being used without any protection and buttons dyed without using an air-extraction system.

Firm 751. Responsibility for these functions had not been assigned to any office; neither did they have written rules for them. Claimed not to experience any safety problems. No periodic fire drills were carried out. They rated their records and procedures in safety and environmental control as about the same as those of other firms in the industry. It was observed that although the plant was crowded there were no flammable materials in view, and no apparent pollution sources.

Firm 1701. No delegation of responsibility in these two areas and no written rules were available. Claimed to have safety posters, but not to carry out fire drills. The General Manager has had prior experience as a safety officer, but claimed not to have had time, due to the pressure of work, to train his workers. They rated their safety and pollution control records and procedures as about the same as those of other firms in the industry.

Firm 161. No assignment of responsibility and no written rules were available. They did not carry out fire drills. Considered their records and procedures in these matters as about the same as those of other firms in the industry.

Firm 271. No responsibility assigned for pollution control and they had no written rules for it. As for safety, they answered that responsibility had been assigned and that they followed Ministry of Labour regulations. There were no fire drills, but smoking was not allowed in the plant. They thought that while their safety record and procedures were about the same as those of other firms in the industry, their pollution control record and procedures rated higher because the cloth-dyeing and printing machinery was relatively new and they had equipment to treat residuals.

Firm 461. No responsibility assigned either for safety or for pollution control; no written rules in either case and no periodic fire drills. Appraised their record and procedures as being in both cases inferior to those of other firms in the industry.

Firm 1121. No office had been assigned responsibility and there were no written rules either for safety or pollution control. Had carried out some fire drills in the past, but discontinued the practice; might do them occasionally in the future. Thought that their safety record and procedures were about the same as those of other firms in the industry, and were unaware of how they rated in pollution control.

Firm 1581. No office was given responsibility either for safety or for pollution control and there were no written rules. They did not carry out fire drills. Considered their safety record and procedures to be better that those of other firms in the industry. General manager said that they have had only a few minor injury cases. Were unaware about how they rated in pollution control.

Firm 1681. While responsibility for safety had not been assigned to any office, they considered their personnel to be well-trained and very conscious about safety. Similarly, they had nobody in charge of pollution control, but paid for refuse removal done by the city, and some materials that could be recycled were sold, in small lengths, to their own staff. No written rules existed, but they considered their record and procedures in both cases to be superior to those of other firms in the industry.

D. Technological functions and organization

1. General

Firm 51. Had patterns and prepared samples and prototypes when a new line or product went into production. Rest of the technical information was tacit and diffused by word of mouth. Thus this plant had little or no technical documentation or files, and no technical offices.

Firm 681. Product specifications, as well as the previously determined production standards set by work study, were included in the product sheet. Also in use were patterns and cut-sheets indicating the various measures. They had a design office staffed by two fashion designers and responding to the Head of Merchandising. Thus this plant could be classified as having little documentation and up to two technical offices.

Firm 751. Only used patterns developed by owner. There were no technical offices or employees assigned to technical functions. It falls then in the category of no documentation, or files, and no technical offices.

Firm 1701. Only had patterns and cut-sheets. There were no technical offices. They employed a designer although her services were presently not utilized because they did subcontracting work not requiring the input of a designer. Falls also in the first category, i.e. plants without technical documentation or files, and with no technical offices.

Firm 161. Had cutting models for sewing, but there was nothing for knitting. No technical offices and no technical personnel. Belongs with plants possessing no documentation, or files, and no technical offices.

Firm 271. Pattern-making was computerized and was used extensively in the plant. Patterns and assembly rules were codified. Had quality control laboratory as well as pilot, or experimental, production line. Claimed to have one quality control staff for every six operators. Had the equivalent of a Product Engineering office where they continuously prepared prototypes. Must follow fashion changes closely and on average had three model changes per year. Plant falls in category with good technical documentation, and three or more technical offices.

Firm 461. The owner designed the patterns and did the cutting himself. No technical offices or technical personnel existed. Falls in the category of plants without documentation or technical offices.

Firm 1121. Technology was tacit and diffused by word of mouth via supervisors and other personnel. No technical documentation, and no technical offices.

Firm 1581. Had a French designer and also quality control personnel, but kept no technical documentation or files.

Firm 1681. Had six quality control personnel, as well as two cutters and one marker. Some designs were provided by the clients. There was little documentation and no technical offices.

2. Product design and product change

Firm 51. Their products were copies of similar imported or domestic products. They had no designer but a merchandiser in their staff. The changes introduced in the products' design had not been significant and were done to suit local tastes, use local materials and make use of available equipment. No new products had been introduced in the last five years, except for changes in styling.

Firm 681. The design of their products arose from: foreign specification, copy of similar imported product, or self-generated specification. Product changes introduced were not very significant. They regularly manufacture three ranges of products: for the summer, Christmas, and winter seasons. Three new ranges had been introduced recently: the sport range, safari range, and clothing for young people. Two years before they had introduced track suites. Designs were copied from a magazine and developed in-house without any outside help.

Firm 751. Since they worked on CMT basis they did not engage in product design or product change.

Firm 1701. Recently have been mostly engaged in subcontracting which did not require that they generate their own product designs.

Firm 161. Design of their products was copied from similar domestically produced products. Changes introduced had not been significant. Product modifications were made to meet market needs. Had not introduced any new products in the last five years.

Firm 271. Designs for ladies' line were self-generated while men's line was created following foreign specifications. Product modifications were made to suit local and international tastes, and also to update designs following fashion trends, but were not significant. New products were introduced in the last five years on two occasions, when they started to export to the US, and to produce pants for the German market. While these new products were developed in-house, they got outside help for the design and testing of prototypes in the men's line for export to the US.

Firm 461. Products' designs were self-generated according to domestic specifications. Fashion trends were followed by examining catalogues from other countries. Changes introduced were not significant. Product modifications were done to suit local tastes, climate and materials, as well as to simplify. They claimed to introduce changes in styling every six months and that these were developed in-house without outside help.

Firm 1121. Not concerned with product design due to the nature of their manufacturing operations.

Firm 1581. Designs were self-generated and also provided by French and German customers' designers. They did not introduce product changes other than to follow changes in fashion and local tastes. They have not introduced new products during the last five years, but changes in models. Their children's clothing line was introduced in 1982.

Firm 1681. Designs were self-generated according to domestic specification or copies of similar imported products. Changes introduced had not been significant. Changes to the products's design were done to suit local tastes, use local materials and to update the design. They had not introduced any new products in the last five years.

3. Process changes

Firm 51. Had not introduced any changes in the production process.

Firm 681. Around 1990 introduced an attachment to the sewing machines to increase their speed. It cost around Z$ 10,000 (US$ 1,540) and the expected benefit was cost reduction. Change was carried out by in-house personnel and the expected benefits seemed to have been attained, albeit data on costs and benefits were not available.

Firm 751. Do subcontracting work and had not introduced any changes in the production process.

Firm 1701. No technical changes in the production process.

Firm 161. No changes had been made to the production process.

Firm 271. Around 1990 they had installed load-limiting equipment to save energy due to its high cost. It was done by outside personnel and the benefits expected were attained. No cost data was available.

Firm 461. No changes in the production process.

Firm 1121. Have slowed down the speed of knitting machinery to adapt to the low quality of local thread. Change was done by in-house personnel and expected benefits were attained in terms of higher quality product.

Firm 1581. No modifications in the production process.

Firm 1681. They introduced new power distribution lines in the plant in 1992. The cost was low: US$ 1,000. Expected benefits were more flexible layout and improved plant safety. Change done with help of local expert and benefits realized as expected, but technical change carried out did not affect directly the production process.

4. Use of technical support services

Firm 51. Have used support services in quality control and repair and maintenance (R&M). Quality testing of a fusable material was carried out for free in Capetown, South Africa, by materials supplier. Once or twice per month they hired the services of a local mechanic to repair machines that break down. He was paid per hour worked. Other assistance was received to check hydraulic and pneumatic systems and to replace some machine parts. This was done once by a local company that specialized in hydraulic and pneumatic systems. This service was also paid by the hour. When asked about technical support services that would be beneficial for their firm if readily available and efficiently delivered, their priority rankings were as follows: 1. quality control testing, 2. training, 3. productivity improvement techniques, 4. technical assistance for trouble-shooting, and 5. process improvement.

Firm 681. They had received technical assistance from South Africa licensor, and used local repair and maintenance services for steam-boiler. The technical assistance consisted of visits by South African technicians, drastically reduced now, plus travelling to South Africa by their personnel to observe the range of new products and fabrics. They paid a royalty fee. The boiler required frequent attention and these services were of course for payment. As to technological services they would like to have, their priorities were: 1. productivity techniques, 2. labour relations, and 3. repair and maintenance.

Firm 751. Only used support services for repair and maintenance when they could not fix the machines in house. They hired services of local company and paid for the service. As to support services they could use, they were only interested in 'productivity improvement techniques'.

Firm 1701. Only made use of R&M support services for their machines. They paid for the services of a local mechanic when used. Their priorities for technological support services would be: 1. productivity improvement techniques, 2. training, 3. repair and maintenance, 4. quality control testing, and 5. product redesign.

Firm 161. Repair and maintenance was the only service they have utilized. It was procured as the need arose and they paid for it. Their priority rankings for the provision of technical support services were: 1. productivity improvement techniques, 2. quality control testing, 3. training, and 4. product redesign.

Firm 271. They have used services for quality control testing and repair and maintenance. The quality control testing was done by an external laboratory. The maintenance referred to support programmes for new computer software provided by the supplier. In both cases the services were paid for. With respect to the supply of technical support services, they were only interested in: 1. training, and 2. quality control testing. The company used to have its own training school which they closed nine years ago for financial reasons. Required to pay 1% of the payroll as a labour training levy, they asked to be allowed to offset the school costs against the levy, but were not allowed to do so. Because of this they closed the school. The textile producers' association was about to open a training college to cater to the industry's training needs and to keep up with international trends. It was expected to provide courses in production management, garment design, plant engineering, etc.

Firm 461. Had not used any technical support services. Their priorities for the supply of such services would be: 1. training, 2. process improvement, 3. repair and maintenance, and 4. product redesign.

Firm 1121. Only used repair and maintenance local service to fix what they could not repair in-house. It referred, in particular, to electrical connections and the provision of castings and some spare parts. They paid for such services. Their priority rankings for technological services were: 1. product redesign, 2. process improvement, 3. productivity improvement techniques, 4. troubleshooting assistance, and 5. pilot plant experimentation. These priorities reflected the general manager's interest at the time of the interview in modifying the manufacturing process in order to produce cotton thread from waste. Specialized equipment would be required for that purpose which he would have liked to buy second-hand from Italy.

Firm 1581. They had received bilateral technical assistance from Germany, and had used local R&M services to rewind burned motors. They paid for the R&M work, but the German technical assistance was free. They helped them with design and marketing to begin exporting. They were very satisfied with this assistance. The General Manager was not interested in any technical support services. She was happy with her operation, and attributed the low labour productivity to low salaries and a lack of commitment by the workers. However, she felt that higher salaries would lead to the firm going out of business.

Firm 1681. Only used repair and maintenance support services provided by a local firm for pay. They minimized the importance of technological support and argued that their main problems were institutional: to obtain access to foreign exchange for imports, and a reliable supply of cloth, elastic and spares. Their rankings for technical support services were: 1. repair and maintenance, 2. process improvement, and 3. tools, dies or fixtures.

On the basis of the rankings shown in Table AII-6, an index was constructed for the technical support services desired by firms visited in the industry assigning the following number of points to each rank: 1 = 5 points, 2 = 4 points, 3 = 3 points, 4 = 2 points and 5 = 1 point. Table AII-7 summarizes the results for the 10 textile and clothing plants interviewed and the top seven technical support services deemed of potential benefit to those firms.

As shown, the industry respondents expressed the greatest interest in receiving technical support for: productivity improvement techniques (six out of 10 respondent firms), training (also six firms), and then quality control, process improvement, and product design, with four firms in each case. The highest rankings were assigned to productivity improvement techniques, followed by training, and then quality control and process improvement.

Table AII-6 Textiles and Clothing. Ranking of desired support services

Firm	*Ranking of support services*				
	1	*2*	*3*	*4*	*5*
51	Q.C.	Training	Productiv.	Troubles.	Proc. imp.
681	Productiv.	Labor rel.	Training	–	–
751	Productiv.	–	–	–	–
1701	Productiv.	Training	R&M	Q.C.	Prod. des.
161	Productiv.	Q.C.	Training	Prod. des.	–
271	Training	Q.C.	–	–	–
461	Training	Proc. imp.	R&M	Prod. des.	–
1121	Prod. des.	Proc. imp.	Productiv.	Troubles.	Pilot pl.
1581	n.a.	n.a.	n.a.	n.a.	n.a.
1681	R&M	Proc. imp.	Tool, dies	–	–

Notes:
n.a.: Did not provide any rankings.
–: Did not provide all five rankings.

Table AII-7 Textiles and Clothing. Overall ranking of support services

Service	*No. of firms ranking it*	*Average score*
Productivity improvement	6	4.33
Training	6	4
Quality control	4	3.75
Process improvement	4	3.25
Product design	4	2.5
Repair and maintenance	3	2.75
Trouble-shooting	2	2
Average	**4.14**	**3.22**

On the whole, these results are very similar to those obtained for the food and beverages industry.

5. Training

A distinction was made in the questionnaire between the training needs of operative workers, technical and middle-level personnel, and professional personnel. In turn, operative workers were potentially divided in three groups: (i) formally trained in the particular job (e.g. via an apprenticeship programme), (ii) with experience acquired in a similar job, and (iii) others, i.e. those without formal training and/or experience.

Firm 51. They did not hire formally-trained operative workers. For those with experience in a similar job, one week of on-the-job training would be normally sufficient to be able to operate the plant's equipment. Those without experience or training would spend about a year training, starting from the lowest job (sweeper) and progressing through the plant. They did not provide training to professional or technical personnel.

Firm 681. Only hired experienced workers sent by the National Employment Council of the Clothing Industry. Their training programme took about five years to reach skilled worker level if starting from the bottom of the skill ladder. More time was required for pattern makers. They do not provide training for professionals. Occasionally provided training opportunities for technicians and middle-level personnel. They had sponsored a technician for a one-year course (after hours) on Work Study for Plant Productivity at Harare Polytechnic. (The complaint was voiced that the firm would not pay for a better course imparted at a private college.)

Firm 751. Did not look for formal education when hiring, insisting instead on intelligence and common sense. On-the-job training generally lasted three–four months depending on the trainer and the candidate's ability. Had two trainees at the time of the interview. No training provided for professionals or technicians.

Firm 1701. No on-the-job training was required for formally trained or experienced workers. For other workers, about three months. The educational qualification required was just primary school completion. No training undertaken for professionals or technical personnel.

Firm 161. Just a month and a half was required to learn to operate the plant's equipment for workers without formal training or experience. Did not hire any professional or technical personnel.

Firm 271. Felt that there were no formally trained workers to be hired and that the existing colleges did not do a good job. That was the justification for the Textile and Clothing Producers Association's decision to set up their own school for the industry. Estimated in 18 months the required on-the-job training for inexperienced workers. They did provide training opportunities for their professional and middle-level technical personnel, of which they claimed to have quite a number. These generally involved training sessions for the upgrading of technical skills, as well as managerial capabilities, imparted by local institutions.

Firm 461. Firm claimed that half its personnel had technical certificates. For formally trained workers they prescribed three–four months' on-the-job training. Experienced workers did not require training, and other workers needed a one-year duration training programme. No training was being provided for professional or technical personnel.

Firm 1121. The plant's jobs were considered of low skills requirements and neither formally trained nor experienced workers would require on-the-job training to operate the plant's equipment. Other workers however would need the training. The General Manager preferred to take in workers that were not too educated. No training was provided for professional or technical personnel.

Firm 1581. Six months' on-the-job training required for formally trained and experienced workers. Other workers would require more depending on the occupation. For machinists, a three-year training period was prescribed. They did not provide training for professional personnel. Technicians received six to twelve months' on-the-floor training.

Firm 1681. Formally trained and experienced workers would not require on-the-job training, but their policy was to only hire inexperienced workers. They hired primary school graduates and trained them for a year. No training was provided for professional or technical personnel.

E. Technological development plans

Only those responses indicating the existence of concrete plans for the future were taken into account.

Firm 51. The firm was planning to expand and to invest in technological upgrading. They had already ordered from Japan 15 mostly sewing and some cutting machines at a cost of Z$ 260,000 (US$ 40,000) essentially destined to expand by 17% their output of trousers. Exports would play the leading role in this expansion although their executives claimed that they contributed only marginally to firm profits. They also expected that the new equipment would permit them to raise productivity and cut costs. Quality changes were not expected. No technical assistance, learning, or organizational changes were envisaged, but they felt that their competitive position would be enhanced as a result of the proposed investment.

Firm 681. Firm was planning to expand exports and to invest in some technological upgrading. The areas selected were improvement of the accounting system and the introduction of computerized cutting. The former was more firm with financing already allocated. The latter would demand about Z$ 250,000 and was more problematic for financial reasons. The motivation for the proposed changes was to improve accounting and the production planning system. The change would involve the acquisition of new equipment (computers) and contracting for the provision of specialized services; a local firm would be hired. It was expected that the offices would operate with higher productivity and lower costs after installation of the proposed new system. It was envisaged that personnel training in computer systems for about six months would be required. Local external assistance would be used to debug and troubleshoot the new processes. The reallocation of some duties as a result of changing from a manual to a computerized system might require some organizational changes but not of a major type. The estimated cost was about Z$ 60,000 (US $ 9,200).

Firm 751. For the time being, and due to market uncertainty, the firm planned to continue as it was, doing CMT and without changes in capacity or technology. Could move to a better room in the same premises.

Firm 1701. While they would like to have more appropriate machines and to establish new lines of production, their aims could not be implemented due to very serious financial difficulties; they seemed to be fighting for survival.

Firm 161. Wishes to buy additional equipment for hemming (overlock), zigzag stitching and button-holes, but had no concrete plans to do so. Were facing building use restrictions and had some machines stored away.

Firm 271. They would like to have more automated equipment to 'de-skill' their operations because they complained about labour's low skill level. However, they had no concrete plans for technological upgrading at interview time and were rather thinking of shrinking due to market difficulties.

Firm 461. Owner had desires and long-term goals for expansion and technological upgrading but no specific and concrete plans. In two–three years would like to add 14 machines at an estimated cost of Z$ 140,000 (US $ 21,500) to be installed in additional space available to the owner. These machines would apply the same technology but run faster. Their acquisition was expected to lead to lower costs, greater productivity and better quality.

Firm 1121. No concrete plans. General Manager would like to be able to buy second-hand reconditioned machinery from Italy, but lacked the financial resources to implement this wish.

Firm 1581. Would like to expand without technical changes but had no concrete plans to do so. Some new machines were acquired three months before the interview.

Firm 1681. No plans for technological improvement. They were looking at increasing productivity while making use of the same technology.

In summary, due to market conditions resulting from liberalization and increased openness to international trade, firms in this industry were facing retrenchment in some cases and in general were not very optimistic about their growth prospects. These circumstances obviously limited their interest and capability for expansion or technological upgrading.

F. Technical skills

The classification of operative workers into skill groups raises serious problems because there was little concordance between the assessments of skill content of different jobs among the various respondents.

Thus we concentrate in this section on professional personnel, i.e. employees holding a university degree, and middle-level technical personnel, defined as those holding a certificate or diploma.

Nevertheless, it should be noted that a variety of post-secondary, vocational courses exist, with different durations and academic requirements, which calls for exercising caution in the way the information about middle-level technical personnel is handled.

Firm 51. They employed two locally trained accountants (one of them is the Managing Director). The technical personnel included two employees with diploma, one an expatriate textile technician (who is the Technical Manager), and another employed in sales. Thus their total professional personnel was two, and middle level technical personnel also two.

Firm 681. The Managing Director held an accounting degree from abroad. Middle-level technical personnel included seven staff with diploma and one with certificate: the Factory Manager held a diploma in textiles, two sales personnel held diplomas, two designers held diplomas, one employee held a diploma on computer-accounting, one employee in personnel also held a diploma, and finally, a work-study person held a certificate. Thus this plant employed one professional staff and eight middle-level technical personnel.

Firm 751. There were no professionals in this plant. The owner held a textile diploma.

Firm 1701. No professionals; two staff with certificates: one in design and one in sales.

Firm 161. No professionals. Owner had a certificate on dress-making.

Firm 271. This plant reported employment of 17 professionals and nine middle-level technical personnel. The professionals included: three engineers trained abroad, two accountants trained locally, four clothing and textile designers trained locally, and eight trained abroad. Middle-level technical personnel included eight staff with textile/clothing diploma, and one with diploma in electronics.

Firm 461. No professionals. Six technical personnel, plus the owner, with textile diplomas.

Firm 1121. Employed one lawyer and one textile mechanic that had not fully completed the certificate requirements.

Firm 1581. Professional personnel consisted of one fine-arts degree holder and two staff with work-study diploma and certificate respectively.

Firm 1681. No professional personnel. The Managing Director (owner-father), held a textile diploma and his wife had a certificate in accounting.

As could be seen comparing with Table A-9, the proportions of both professional personnel and middle-level technical personnel in total employment are smaller in the textile and clothing industry than in the food and beverages industry. The difference is particularly striking for the proportion of middle-level technical personnel: 1.3% against 7.0%. This corroborates the hypotheses about relative skill intensity of the industries included in the study.

Table AII-8 Textiles and Clothing. Professional and technical personnel

Firm	*Profess.*	*Diploma*	*Certifi.*	*Employ.*	*% Profes.*	*% D + C*
51	2	2	0	480	0.4	0.4
681	1	7	1	231	0.4	3.5
751	0	1	0	11	0.0	9.1
1701	0	0	2	95	0.0	2.1
161	0	0	1	2	0.0	50.0
271	17	9	0	1554	1.1	0.6
461	0	0	7	13	0.0	53.8
1121	1	0	1	47	2.1	2.1
1581	1	1	1	156	0.6	1.3
1681	0	1	1	49	0.0	4.1
Total	**22**	**21**	**14**	**2638**	**0.8**	**1.3**

Notes:
D: Diploma
C: Certificate

Appendix D: Firm Interviews by Sector: Wood and Furniture (ISIC 33)

Of the 10 firms surveyed, seven manufactured only furniture, one produced a specialty product: billiard tables, one besides wood furniture also manufactured some metal furniture, and one firm in addition to furniture also produced boxes and wood trusses. Most firms manufacture home furniture with a couple of them also producing office furniture.

A. Background information

1. General

As shown in Table AIII-1, firms in the sample vary significantly in size (measured by employment), output per worker, ownership, and ethnic origin.

The plant producing industrial products (timber and boxes) in addition to furniture, had by far the highest output per worker. Larger firms tended to have

Table AIII-1 Wood and Furniture. Basic characteristics of firms interviewed

Firm	*Product*	*Employment*	*Ownership/ethnic*	*Location*	*Sales/worker US$*
41	Cabinets/ Furniture	4	African	Harare?	n.a.
131	Billiard Tables	14	European	Harare	8,800
231	Furniture/ Box., Wood	197	European	Harare	39,000
331	Furniture	195	European	Harare	n.a.
1191	Furniture	115	European	Bulawayo	12,700
1471	Wood/Metal Furniture	77	European	Bulawayo	11,000
471	Furniture	236	European	Harare	5,600
1771	Home/Office Furniture	107	African Cooper.	Harare	n.a.
1861	Home/Office Furniture	220	European	Harare	n.a.
1921	Furniture	358	Holding	Harare	18,600

also higher output per worker probably on account of their greater mechanization and investment in machinery, although there were exceptions to this pattern. It should be noted that several interviewees failed to provide turnover figures.

2. Market orientation

With about half of its area covered by forests (see Chapter 1), Zimbabwe has a plentiful supply of raw materials for the manufacture of furniture and other wood products. This natural resource endowment confers on the country a potential comparative advantage which, if realized, would be reflected *inter alia* in the proportion of the output of this industry exported.

Table AIII-2 shows that three of the 10 firms in our sample did not export at all, while one firm exported 90% of its output. Two firms exported more than 30% and two others 20% of their sales. The large exporters were also the largest firms in terms of sales. Home furniture products were the largest item exported to Europe and to neighbouring countries. Exports were favoured with a subsidy, tax drawback, and better access to foreign exchange. Domestic prices, for similar products, tended to be higher than export prices.

3. Organization

The survey inquired about the presence of professional management, the kind of organizational structure adopted, and the existence of certain formally assigned managerial functions.

As shown in Table AIII-3, a majority of firms did not employ professional management, relying instead on family or entrepreneurial self-management. Also there was little prevalence (three firms in 10) of several tiers structure in the organizations adopted. In two firms there was direct supervision by top management of the workers, and in five others only one tier level delegation of managerial authority existed. The larger firms tended to have professional management and one or more tiers of functional management.

Table AIII-2 Wood and Furniture. Proportion of output exported

Firm	*Exports $US*	*Sales $US*	*% Exported*
41	0.0	n.a.	0.0
131	6,150	123,000	5.0
231	3,000,000	7,700,000	39.0
331	n.a.	n.a.	10.0
1191	480,000	1,500,000	32.0
1471	0.0	850,000	0.0
471	260,000	1,300,000	20.0
1771	0.0	n.a.	0.0
1861	n.a.	n.a.	90.0
1921	1,340,000	6,700,000	20.0

Table AIII-3 Wood and Furniture. Organizational structure

Firm	*Profess. manag.*		*Organizational structure*				
	Yes	*No*	*Direct (1)*	*Super. (2)*	*1 Tier (3)*	*Tiers (4)*	*Other*
41		X	X				
131		X	X				
231	X					X	
331		X			X		
1191	X					X	
1471		X			X		
471		X			X		
1771		X			X		Coop.
1861	X				X		
1921	X					X	

B. Entrepreneurial history and acquisition of technology

1. Origin of the firm and initial technology

Firm 41. Owner (black), and three workers who all claimed to have wood-working certificates, manufactured and repair kitchen cabinets, coffee tables and wardrobes. Operated in a shed without power or water, and only using hand tools. Due to its limited equipment produced with very low productivity.

Firm 131. Owner started production of billiard tables as a backyard hobby activity when still employed as chief executive at a printing firm. At the time of interview had worked full-time on this line of business for some three years of which only the last in the new premises. Used granite slabs, cloth plus wood for the manufacture of the tables that must meet very high quality standards in terms of horizontal levelling and smoothness of the top, plus the hand-done wood-finishing. Bought granite and cloth for the tops, and obtained free hard-woods from an 'abandoned' mine. Owner was very proud of his level of quality and craftsmanship.

Firm 231. Group associated to tobacco and wood sales bought the firm in 1991. Strong push for exports to Europe of pine wood furniture (mostly beds). Also sold boxes and wood for construction locally.

Firm 331. Begun as a very small shop that the owner bought in 1967, and expanded. Owner and managing director's family had a retail furniture store; he grew up in it and learned to appreciate his job. Claimed to love furniture. Manufactured up-scale furniture, supplying, for example, three of main local hotels.

Firm 1191. Plant was established in the 1930s and changed hands several times. Present owner bought it in 1979. There have been no significant plant investments since the 1940s. A new manager was hired recently to change and modernize the plant. Labour force was also quite old and many were scheduled for retirement. Had a small kiln to dry the wood, an upholstery shop, and also manufactured metal spring mattresses for their beds and for separate sale.

Exported about one-third of their output, particularly dining-room sets, to South Africa and Botswana.

Firm 1471. Entrepreneur without experience in the industry, bought a controlling interest in the firm in 1991 and has expanded output and increased productivity substantially since that time. Produced furniture for the middle income range and only for the domestic market.

Firm 471. Sephardic Jewish entrepreneur, born in Zaire from family that emigrated from Rhodes island in Greece. Had experience in clothing manufacturing and bought this firm in August, 1973. It was small and undergoing difficult times. It had a few basic machines and employed 12–15 workers. His uncle who had experience in the furniture business said that it was worthwhile. Improved operations from the outset and had been expanding since. Initially aimed at the lower end of income market having moved more recently to the middle-income range. Moved to the present premises in 1980–81 and introduced new lines of furniture. Planned to move to new premises shortly and to further expand production and exports.

Firm 1771. Former employees, all carpenters, of another firm that closed decided to open shop as a cooperative after contributing monthly to raise capital. Ten of initial 26 employees founded in 1986 the present company. Bought small plot of land from the municipality and initially used manual tools and invested their pension funds to start new production operating on open air. In 1987 obtained loan from SEDCO, that also assisted with preparation of a feasibility study. International donors and NGOs also provided help in the form of machinery and working capital. Planned to initiate exports in the near future.

Firm 1861. Family firm started in 1960 by Scottish parents, now retired, of present owner. Father was shop-fitter and mother accountant. Before UDI they worked in shop-fitting, but after UDI started to manufacture Danish-type furniture. In 1981 started exports to South Africa. Were presently practically fully oriented towards exports: hardwood furniture to South Africa and pine-wood furniture to Europe (United Kingdom). Firm had grown substantially but they still expected to increase output.

Firm 1921. Holding group with retail and manufacturing divisions bought bankrupt enterprise (the same from which emerged the cooperative that constituted Firm 1771) in 1986. Present operation included three divisions: pine-wood furniture, hardwood furniture and bedding. Hardwood furniture manufacture was mostly manual while the pine furniture was mostly knock-down and more mechanized. They got two IFC loans for the acquisition of new machinery. Pine-wood furniture was all exported, while the hardwood furniture was 25–30% exported, and only some 10% of the bedding was exported.

2. Sources and characteristics of equipment

Because it was not feasible to carry out a complete census of all available equipment in each of the plants visited, the questionnaire included questions about the three principal pieces of equipment in each location.

Table AIII-4 summarizes the information obtained about origin, age and whether the machines were bought new or used.

As shown in Table AIII-4, the main source of wood-working machinery has been the United Kingdom, followed by Germany. Zimbabwe and Taiwan were also mentioned in a couple of cases.

While clearly a vast majority of the machinery seemed to have been bought originally new, there were three factories were the principal equipment was between 20 and 60 or more years old. Three other factories seemed to have quite new principal equipment – all less than 10 years old.

Table AIII-4 Wood and Furniture. Information about key pieces of equipment

Firm	*Origin*	*Age (years)*	*Bought new or used?*	
			New	*Used*
41	Zimbabwe	12	X	
	U.K.	3	X	
131	Taiwan	10	X	
	Taiwan	8	X	
231	U.K.	5–25		X
	U.K.	10	X	X
	U.K.	<5	X	
331	Germany	30		X
	Italy	30		X
1191	U.K.	>60	X?	
	U.K.	>60	X?	
	U.K.	>60	X?	
1471	Zimbabwe	20	X	
	Various	20–40	X?	
471	Germany	5	X	
	U.K.	10		X
	U.K.	12	X	
1771	U.K.	5	X	
	U.K.	5	X	
	Germany	9	X	
1861	France	4	X	
	U.K.	1		X
	Germany	3	X	
1921	U.K	30	X	
	Holland	30	X	
	U.K.	30	X	

3. Use of foreign technology

The firms interviewed were asked if they had a foreign licence or technical assistance agreement and whether they received help for commissioning the plant or installing the equipment.

Firm 231. Had a technical assistance arrangement to make trusses with CAD. They paid a fee for the service.

Firm 1771. German NGO, provided volunteer services of an engineer who stayed for three years. They paid his rent and the NGO paid his salary. An engineering company assisted in the commissioning of the plant.

Firm 1861. Technical personnel from supplier came to instal equipment, and firm personnel went to South Africa to train for a week at Siemens on how to operate numerically-controlled router. For a new multiple-head lathe they brought a technician from the US for a year to teach tool preparation and machine set up.

Firm 1921. To instal new equipment, technical personnel from supplier came and their personnel received two weeks of training.

Thus the wood-working firms interviewed did not use licences and/or formal technical assistance agreements to obtain their technology, but several made use of external services to commission their new equipment and to train their personnel on how to operate it.

C. Operating capabilities

1. General

Firm 41. Minimal production facility in shed with no floor, electricity, water or sewage. Some work was done outside the shed. The equipment consisted of

Table AIII-5 Wood and Furniture. External technical assistance

Firm	*Licence*		*Technical assistance*		*Outside help*	
	Yes	*No*	*Yes*	*No*	*Yes*	*No*
41		X		X		X
131		X		X		X
231		X	X			
331		X		X		X
1191		X		X		X
1471		X		X		X
471		X		X		X
1771		X	X		X	
1861		X		X	X	
1921		X		X	X	

hand-powered simple woodworking tools. Worked one shift claiming full capacity utilization and a backlog of orders on hand.

Firm 131. Unorganized but clean and well-lighted facility with few, relatively new, machines. Worked one shift, and claimed 100% capacity utilization. One machine was in repair shop for long time. Kept no spare parts.

Firm 231. Good facilities but the plant appeared disorganized because they were preparing for the addition of new machines and a substantial change in layout. Part of the equipment was old. Planned to expand production with the introduction of new machinery. Temporarily worked two and three shifts in some sections to meet demand and because of unbalanced process. Night-shift output was somewhat lower. After expansion would revert to one shift which seemed to be the industry standard.

Firm 331. New plant, clean, well-lighted and appeared well organized, however some lumber was stored in wrong place, and dust container was open. Most machinery was quite old and was acquired used. Some problems with repairing one machine. Worked one shift and claimed to be operating at full capacity, which interviewers doubted.

Firm 1191. Very large old facilities, well-lighted, but poorly kept. Some sections of the plant were messy and with too much dust. Did not appear to be very safe. Worked one shift and claimed 100% capacity utilization, although there were idle machines and frequent breakdowns because of the machines' age (more than 60 years).

Firm 1471. Modern, spacious, well-organized and well-lighted building, but output had grown too fast and the place was getting crowded and with too much dust in the air which might make it unsafe. Needed storage space for finished products. Had plans to expand building. Most equipment was old with a number of machines over 25 years old. Worked one shift. Lots of activity in plant, no idle machines, and workers appeared to be very busy. Capacity utilization was probably over 100% as they were working six, instead of the normal five, days a week.

Firm 471. Very crowded and dirty plant. Poor dust collection made it dangerous. Poor storage facilities. Finished product and incoming wood were unprotected. Planning to move to new facility. Average age of equipment was 8–10 years. Claimed to work at 100% rate of capacity utilization. They were working overtime during the week and also on Saturdays. No idle machines observed.

Firm 1771. Congested plant with poor lighting and poor ventilation. Most workers were not using their protective goggles. Very poor storage for raw materials and not enough space for finished products. Dust collection was good but painting areas were not properly ventilated. Equipment was mostly new and safety warning posters were clearly displayed. Claimed to work one shift at 90% rate of capacity utilization. Due to lack of space, some wood-working and upholstery machines were not being utilized.

Firm 1861. Appropriate premises, well lighted but dust not well extracted and spray painting area was not well ventilated. Raw timber not well stored. Equipment was relatively new. Worked one shift at 60% rate of capacity utilization because of unbalances in production line which led to some machines being idle.

Firm 1921. Well-adapted premises, well ventilated and with lots of light except for spraying areas. Majority of equipment was quite old (30 years). Worked one shift at 55% capacity utilization in the pine-furniture section and at 80% in the hardwood furniture section. Some machines were idle because of inadequate quality of timber received for the pine-furniture manufacture. Planning to expand pine production line into two shifts to meet export demand.

2. Troubleshooting

Firm 41. Nothing to report given the primitive, manual, nature of its equipment.

Firm 131. Experienced limited breakdown problems that were solved by in-house personnel and with some outside local help.

Firm 231. The technical problems experienced during the last two years included: equipment breakdowns, serious quality control problems, bottlenecks and pollution control difficulties. Machinery breakdowns were solved by in-house maintenance department with some outside local help. Except for electric motors that were sent out to be fixed. Carried sufficient spare parts for all machines. Main quality problems were in handling and finishing. Loose assembly and poor spraying and machining were less important. Quality control was brought to the production level making plant manager responsible. They had also set up a quality control team and made each worker responsible for his area. The existing plant imbalance was being corrected by new plant investment. The pollution problem arose because the cyclone was leaking and dust was everywhere. They hoped to solve the problem soon.

Firm 331. Main technical problems reported were: machine breakdowns, quality control and bottlenecks. Machine parts were some times not available and they have had to manufacture spares in outside local shops. One machine tended to break down and was being used below capacity. There was no back-up service from supplier. They reworked components that were not up to standards. Sometimes the quality problem was caused by the design requested by customers. Bottlenecks due to differences in output among various machines have not been solved yet.

Firm 1191. Equipment breakdowns was a big problem due to the very old age of the machines. Quality problems were blamed on poorly trained work-force. Products were inspected visually and had to be checked frequently. The old machines were unsafe because they did not have appropriate guards. They needed better dust extraction equipment to avoid pollution and safety problem.

Firm 1471. They experienced machine breakdowns but fixed most problems in house. Rewiring of motors was done outside. Aware of the high level of saw dust in the air, but were putting other investment ahead from spending in the necessary air extraction equipment.

Firm 471. Depending on product being manufactured they experienced bottlenecks in sand-paper section and machine shop. They worked overtime to pair up.

Firm 1771. Machine breakdowns were remedied with outside local help. Imbalance in production due to lack of sufficient space and poor transport facilities. Problem had not been solved yet. Would like to construct bigger plant near present premises. Pending financing availability, planned to buy a bigger delivery lorry (truck). They also needed new lacquer spraying equipment with water curtain, to avoid pollution problems. (It was not clear to the interviewers to what extent these were 'legitimate' statements about existing problems or the expression of wishes for additional international assistance.)

Firm 1861. Machine breakdowns were frequent and they needed to keep spares. During the survey they needed bearings that had to be custom made up locally. Existing bottlenecks and pollution problems had not been solved yet. Spraying operation and dust extraction system needed improvement.

Firm 1921. Bottleneck problem not solved yet, but they were planning changes in the production layout.

3. Quality control

Firm 41. Materials, parts, process and final product were subject to visual, practical inspection. No standards and no routine for sample-testing during production process. Considered their quality to be equal to that of other local manufacturers. Had no idea about reject levels but thought that customers were satisfied. There were no methods to keep track of client complaints and no system for statistical quality control (SQC).

Firm 131. Final product, processes and raw materials and component purchases underwent visual, practical inspection. No standards and no sampling inspection routine. However, owner-manager claimed to supervise continuously the production of his expensive product (billiards table). Affirmed to have no local competition and that his quality was comparable to that of imports. Asserted to have had the last three years an internal reject rate below 0.25%. Maintained direct contact with customers to track complaints. No SQC system.

Firm 231. Materials and components in process, processes and the final product, were all visually inspected. Timber was bought dry and the level of humidity had to be checked. The timber they got was up to Standards Association of Zimbabwe (SAZ) norms. They used international safety standards for their exports. In particular, had to meet child safety requirements for beds in European markets. Did not have a sampling inspection routine during the production process, but did some random checking. Claimed to have much higher quality than other local producers, and even higher, and coming up, than imports. Last year they had a 1.6% replacement rate which they expected to drive down to 0.5% in six months. Three years ago the rate was 3%. So there seemed to be an improvement trend. The production manager was in charge of quality control and seemed to be a good troubleshooter. He had established several check-points in the furniture line and low-quality finishes were taken care of before the product was ready to be packed and shipped out. They maintained direct, and through their sales force, contact with their customers who came to visit them. They also travelled to Europe twice a year. There was no SQC system.

Firm 331. There was visual, practical quality control of final product, raw materials and other purchases as well as materials and components in process, but no controls of production process(es). No norms or standards were in use. Considered the quality of their products to be higher than that of local competitors and equal to that of imports. Quality was improving. Faulty components were detected and removed before the final product was put together. Kept track of customer complaints by direct contact with their clients. There was no SQC system.

Firm 1191. Raw materials, parts and components, processes and final product were subject to visual, practical, quality control. For control of humidity in lumber the general manager relied on his experience and just examined the wood. While they produced some high-quality products, no standards were in use, there was no testing of samples, and the plant operated like a large artisan shop with the control being done by experienced workers and the general manager. Considered their quality to be higher than that of local competitors and equal to that of imports. Rejects were quite high three years ago, but less than 1% last year. They were improving and the new manager was striving for excellence thereby eliminating unnecessary rejects. Kept in touch with customers via the sales force, and had established a good relationship with their buyer in South Africa. No SQC system.

Firm 1471. All inspections were visual, practical and no standards or norms were in use. They checked every piece produced. General manager believed that he could compete in South Africa's lower tier market, but not in Europe. Held his quality to be higher than that of local competitors. Last year rejects were minimal because they took care of imperfections in the production line. New manager had improved output and quality. Clients' complaints were addressed by the general manager and the sales force. No system of SQC.

Firm 471. There was no control of raw material purchases and other inspections were all visual and practical. No norms were in use and no sampling routine during the production process, but they did examine the final product. Considered their quality to be higher than that of local competitors and equal to that of imports. Rejects were about 2–3% last year and the same three years ago. They got the product back when there was a complaint and had direct contact with customer. No SQC.

Firm 1771. Timber was inspected in situ by purchaser and inspectors did visual inspection at the plant. Everything was inspected visually and in practical ways. No use of standards or norms. No sampling routine for testing during production process. Considered their quality to be similar to that of both local and foreign competitors, but they were not exporting yet. Last year they had 1% rejects due to upholstery problem. That section had been greatly improved. They had sent a questionnaire to retailers to obtain their feedback. For major complaints they used the telephone service. No SQC.

Firm 1861. Final product was inspected 100%, visually. Timber purchases were humidity-controlled with hygrometer. Processes and components were also subject to visual, practical inspection. They did not use norms but their exports were subject to British buyer's specifications. No sampling routine for process inspection. Believed their quality to be higher than that of local competitors, but could not judge how it compared with imports. Internal rejects last year were 2–3%, and three years before greater than 3%. They were improving quality by providing better information to the supervisors. Clients' complaints were addressed via the sales force and by direct contact with customers. They did not have a SQC system, but had contracted with local office of major international accounting firm to implement the Kawasaki Production System (KPS).

Firm 1921. Used norms for humidity control in timber. Final product and processes were subject to visual, practical controls. Must meet EEC standard for pine products as a requirement of buyer. Adopted in July of 1993. They used sampling to inspect units produced. They believed their furniture to be the best in the country. There was another producer who was deemed to be better for cabinet making. They believed to be at par with other firms exporting. The existing 40% duty on imports gave them a competitive edge which they need. Its removal would cause problems. Last year they had zero rejects in hardwood furniture but about 15% in pine products. Wood quality was the problem and quality control was only done at the end. Seconds were sold to a third party. Expected to tighten quality control procedures and to reduce the reject rate. Complaints were dealt by direct contact with customers, both retailers and importers (for pine products), and via the sales force. There was no SQC system.

4. Repair and maintenance (R&M)

Firm 41. Minimal R&M needs because they use hand, non-powered, tools.

Firm 131. Weekly servicing and parts replacement when needed; son is in charge of R&M. Outside help required for some machines. Claimed capacity to strip most machines and replace parts as needed.

Firm 231. Major overhaul at time of yearly vacation when plant shuts down. Outside help needed only to fix electric motors. Breakdowns were infrequent and they quickly took care of them. Machine shop was under supervision of purchasing and maintenance manager. Able to strip most machines to replace parts as required.

Firm 331. Machine breakdowns were frequent due to very old equipment. Management laid fault with poorly trained work-force. Had R&M shop with mechanic in charge, but no maintenance routine. Could fix most machines with some requiring outside help.

Firm 1191. Machines were very, very, old and broke down frequently. No in-house R&M facility and no maintenance routine. Outside help was required to overhaul machines and to make spare parts; they could service, oil, etc. R&M was the responsibility of factory management.

Firm 1471. Annual shutdown and overhaul of machinery. Every weak experienced minor problems. Parts were replaced during overhaul and when they failed. Machine shop and maintenance department had the capacity to strip most machines and to replace parts if necessary. Outside help needed mostly to fix electric motors.

Firm 471. Machine shop manager was in charge of R&M. There was no maintenance routine and machines were checked only when something broke down. Oiling and other servicing were done at different frequencies according to type of equipment. Outside help was rarely required as they had the capacity to service and to strip most machines to replace parts as needed.

Firm 1771. There was no maintenance routine but a yearly overhaul at year's end. Oiling and other servicing were done weekly with no established frequency for parts replacement. No outside help required for most R&M tasks but a local British firm came every 400 hours of operation to service compressed air equipment. Breakdowns were infrequent. There were two mechanics with three years of training each. Production manager and foremen were responsible for R&M and alerted to problems by the workers. Claimed the capacity to strip most machines to replace parts when required.

Firm 1861. Had R&M shop with four workers. Plant manager was responsible for R&M. There was no established routine for preventive maintenance. Oiling and other servicing were carried out weekly and no frequency had been established for parts replacement. Breakdowns were infrequent and local services were sometimes required. R&M capacity limited to simple servicing, oiling, etc.

Firm 1921. Factory maintenance foreman reported to Bedding section manager, and four mechanics reported to him. There was no established maintenance routine. Breakdowns were frequent but minor. R&M capacity limited to simple servicing, oiling, etc.

5. Industrial engineering

Firm 41. Did not carry out any organized effort to improve productivity, and did not know how his output per worker compared with those of other competing firms. There was no advanced scheduling of production. Bought lumber weekly week and kept inventory of raw materials.

Firm 131. Owner considered his product (billiard tables) to be individually crafted and was unaware of how his productivity compared with that of other local or foreign competitors. Tried to plan production one month in advance at least. No formal inventory controls and no computers were in use. Plant layout was reexamined only on occasion of expansion or machine replacements. Thought about changing it when some big machines were returned from repair shop.

Firm 231. Production Manager was making improvements in the production process to cut unnecessary operations which had resulted in time savings and the reduction of material handling. Believed their productivity to be higher than that of competing local firms, and similar to that of foreign companies (had South African firms in mind). Tried to plan production three to four months in advance, but also aimed to retain flexibility to fulfil short-notice orders from the best customers. Kept a tight delivery programme for exports. Had Cardex system of inventory control for raw materials, work in progress and the final product. Took stock once a month and tried to maintain 'strategic' levels of stocks. Plant layout had changed recently because of several expansions. Unlikely to modify it otherwise.

Firm 331. Believed that their productivity was similar to that of local competitors but lower than that of producers in South Africa. Tried to plan production four months in advance, but when there were not sufficient orders switched to a month to month system. Used Cardex system for inventory control of raw materials and final product. No computers in use. Plant layout was changed in early 1992 when they moved to the presently occupied premises. Would re-examine it only on occasion of plant expansion or replacement of equipment.

Firm 1191. General Manager set production standards, in conjunction with Plant Manager and Factory Supervisors, based on his own experience. Some time ago a Japanese expert conducted a time-and-motion study, but it was not being applied. Manager was willing to sacrifice quantity for quality, and complained bitterly about the poor education of the work-force, but their machinery was also very old. Admitted that their productivity was lower than that of local as well as foreign competitors. Tried to schedule production four to six months in advance, e.g., Christmas sales plans were finalized in August. Kept computerized inventory control for all items that lend themselves to it. Used several IBM clone PC machines. Plant layout reexamined only if plant expansion under consideration or machines were to be replaced.

Firm 1471. General Manager had training in time and motion studies but no experience in wood-working. Daily production goals had been established and workers received incentive payments if targets were surpassed. It was expected that workers would receive five to six extra weekly salaries during the year. Production Manager had been sent for training in supervisory management to the UK. They had increased output substantially with the same labour force and equipment. Estimated that their productivity was higher than that of other local firms but did not know how it compared with that of foreign competitors. Production was planned on basis of delivery dates established for orders and they tried to plan one to one and half months in advance. Inventory of raw materials, work in progress and final product, controlled with Cardex system and daily output targets. Layout was being re-examined in view of output growth which required that a few changes be made. Planned to expand their premises.

Firm 471. The Managing Director set production standards based on his experience. A bonus system was in place for upholsterers and carpenters who do not operate machines that determine their own output. Because they specialized and have standardized some production phases, they believed that their productivity was higher than that of local competitors, but were not aware about the performance of foreign firms. They planned forward for at least a week. From the time they got an order, to filling it, only spent two to three days; in this way they cut the need for stocks. There was no formal inventory control system; they simply checked stocks visually. They needed more space and planned to move to new premises, at which time they would basically retain the same layout, but with some changes.

Firm 1771. Management set daily production standards which were handed over to the foremen for execution. Estimated to have higher productivity than competing local firms, but to be below the level of foreign competitors. Planned production one month ahead and tried to avoid re-scheduling of orders because of the disruption it produces in the plant. Had formal inventory control of raw materials and work in progress. Final product was not kept in premises. No use of computers and layout would be re-examined only in case of plant expansion or replacement of equipment.

Firm 1861. New General Manager was in process of installing KPS. There was one locally trained technician with experience in time-and-motion study who would be setting standards and in the future monitor productivity. Unaware of how their productivity compared with that of other local or foreign firms. Planned production one month ahead with the plan further broken down by weeks. Inventory control was by Cardex system and they were transferring to computer system which was only partially in use. Had an internal network of eight PCs and used a manufacturing software package integrating accounting, payroll, sales, and the monitoring of stocks and production planning. Would re-examine layout on occasion of plant expansion or equipment replacement. Last change was four years ago when they brought in new machines and improved the dust extraction system.

Firm 1921. No production standards were in use. Gave confused reply to question about productivity comparison. Tried to schedule production three months in advance, but had been experiencing backlog. There was no stock of final product. Cardex system in use for raw material stocks. No computers in use. Claimed to assess the plant layout continuously with a view towards its improvement. Last change was done when they brought in new machines.

6. Industrial safety and pollution control

Firm 41. Minimum equipment in use justified only limited concern either about safety or pollution control. Estimated that their safety record and procedures were about the same as those of comparable firms, and were unaware of how they compared with regard to pollution control.

Firm 131. Only had safety warning posters. No fire drills and no responsibility assignment either for safety or for pollution control. In both fields rated their records and procedures below those of other firms in the industry.

Firm 231. There was a health-and-safety workers committee, and they claimed to have written rules for safety and environmental control. Did not have periodic fire drills but were installing fire protection sprinkler system, and four workers had been trained in the use of fire-fighting equipment. Workers were seen using protective masks and earplugs. Held their safety and pollution control records and procedures to be above those of other firms in the industry.

Firm 331. Responsibility for industrial safety had been assigned but not so for pollution control. There were no written rules either for safety or for environmental control. No fire drills were carried out. However, it was a new manufacturing facility with good dust removal facilities, and they claimed to have been problem-free. Nevertheless, the interviewers observed that paint-shop workers were not using the protective masks available, and not all painting fumes were extracted to avoid complaints from neighbours.

Firm 1191. A nurse visited the plant weekly to talk about workers' problems and teach them how to use safety equipment. No responsibility had been assigned for, and there were no written rules about, environmental control. There were rules about the wearing of overalls and protective masks. No fire drills were done. They have had one serious accident about four months before, and averaged about three accidents per year. This record seemed to be quite high, but respondent blamed the machines, and believed that their records and procedures, in both safety and pollution control, were better than those of other firms in the industry.

Firm 1471. While responsibility for safety had been assigned, workers seemed to disregard instructions and were not punished for it. No responsibility assigned for pollution control, and air extraction system was badly needed. No written rules, only some posters, for safety, and no written rules for environmental control. No fire drills. Rated at about the same level as those of other firms in the industry, their safety record and procedures, and lower in the case of pollution control.

Firm 471. Each supervisor was responsible for safety in his section. Responsibility had not been assigned for pollution control. Safety gear had to be used in painting section. No fire drills and no sprinkler system, but fire brigade visited the plant regularly. Safety situation looked pretty bad compared to other firms due to severe congestion. Pollution control record and procedures were estimated to be about the same as in other firms in the industry.

Firm 1771. Nobody was assigned responsibility for industrial safety or pollution control. No written rules for either safety or environmental control, only posters with safety warnings. Workers did not use protective gear assigned to them. Painting and spraying sections were problem areas because of lack of proper ventilation. Unsafe storage of raw materials observed. There was an assigned fire-fighter in each department, but no periodic fire drills. Held that their records and procedures for both safety and pollution control were comparable to those of other firms in the industry.

Firm 1861. Production Manager was responsible for industrial safety, but nobody was in charge of pollution control. They intended to enhance extraction of dust throughout the plant and of fumes in spray painting section. Claimed to have written rules for safety but not for environmental control. Had hired a private firm to carry out periodic fire drills. Rated their safety and pollution control, records and procedures at about the same level as those of other firms in the industry.

Firm 1921. Each supervisor was responsible for safety and they had some posters displayed, but nobody was responsible for environmental control. No written rules either for safety or for pollution control. Private company was hired to conduct periodic fire drills. Had a full-time clinical nurse, which is seldom found in companies in the industry. Held that their record and procedures should be rated higher than those of other firms in the industry as regards safety, but lower in the case of pollution control due to the situation in paint spraying.

D. Technological functions and organization

1. General

Firm 41. Very primitive operation with no permanent facilities and manual powered tools. No technical documentation, offices or technical employees.

Firm 131. No technical documentation available; it was all in the owner's head. No technical offices or employees.

Firm 231. Claimed to have cutting lists for the various products, as well as manuals and specifications for the operation of machines. No technical offices, but for the manufacture of trusses, computer was used to make drawings of the various cuts and strength estimates.

Firm 331. Products were codified and they claimed to have specifications and blueprints for their manufacture. Under Managing Director's leadership they had designed and produced new models and products.

Firm 1191. Technical information mostly tacit with a few blueprints for each product-model. Design of new products, such as a TV cabinet, done following General Manager's instructions and sketches. He read catalogues, visited fairs, etc.

Firm 1471. Had cut lists and drawings. New models were tried out in the plant. No offices or personnel assigned to technical functions.

Firm 471. Had catalogue and cut sheets for different sizes. In upholstery there were templates. No technical offices and no technical personnel.

Firm 1771. Designs, in the form of simple sketches with measures, were prepared by Production Manager's assistant. No technical offices or other documentation.

Firm 1861. A draftsman, hired six months before, prepared full drawings of parts which were sent to the shop with the production order. They also had cutting lists and samples. The design function was supervised by the General Manager.

Firm 1921. For hardwood furniture they prepared sketches copying from magazines. For the pine line they had a cut list for each product and each particular section. Drawings were provided by British customers. They also made their own designs now.

2. Product design and product change

Firm 41. Designs originated in domestic specification or were copies of similar imported products. Used a catalogue prepared in Yemen and given to him by the country's President. Only minor modifications introduced to meet market needs. No new products introduced in the last five years.

Firm 131. Customers provided specifications but billiard tables have standard measurements. Owner's wife designed the legs and other accessory parts. Changes introduced were not very significant and made to satisfy customers. No new products introduced in the last five years.

Firm 231. Designs for exports were provided by importing customer. Occasionally generated design for domestic products. Changes introduced have been not very significant and done to suit local tastes, adapt to the use of available equipment, and to update the design. No new products were introduced in the last five years, but they claimed to have been introducing new models continuously. For example, for new bed models, customers provided the design and the testing was done at the University of Zimbabwe.

Firm 331. Designs were self-generated according to customer's request, or copied from foreign models seen in overseas fairs. In some cases introduced significant changes to the product, in others they were not substantial. Product modifications done for a combination of reasons: to suit local tastes, adapt to the use of local materials, to simplify product, to adapt to available equipment, to update the design, to differentiate the product, and to incorporate new functions. Claimed to have introduced new products on a continuous basis.

Firm 1191. Product designs were copied from imported catalogue or self-generated according to domestic specifications. Changes introduced were not very significant, as for example in the case of TV cabinets developed in-house. Product modifications done because their old machines could not reproduce all details of some high-quality products. Introduced new product last year – TV cabinets, which were developed in-house without any outside help either for the design or testing of prototypes.

Firm 1471. Designs were self-generated, following domestic specification, or copies of similar imports. Owner admitted lack of knowledge in design and got some help from retired designer friend. Product modifications have been minor, and done to update design. They have introduced new products last year and during the last five years. Recently, a client asked to have some metal lawn furniture which they prepared to order without any outside help. They have now decided to establish a new product line of metal furniture.

Firm 471. Got ideas from abroad but did not design a wholly new product. For the exports to South Africa, they did some modifications, and took samples to show. Buyers also came and saw samples. Product modifications were not very significant. They did them to: simplify the product, update the design, and adopted new materials and better finishing for the export market. Modifications were done in-house without outside help.

Firm 1771. Designs have been self-generated as well as copies of similar domestically produced product. Recent new product designs have been an executive chair and TV cabinet. The changes introduced for example in the TV cabinet were minor (put wheels, and changed doors). They have introduced new products during last year: executive chairs and office furniture. Changes and new designs were developed in-house without outside help. The changes were done to better meet local tastes, use local materials, and to incorporate new functions (add mobility in the case of TV cabinet).

Firm 1861. In recent years they have developed a line of office furniture. Received samples from South African buyer and copied the designs by 'reverse engineering'. They did not use any other outside help. Changes made were not significant, done to be able to use the existing equipment and to save in wood costs.

Firm 1921. They have developed a new line of pine furniture last year on the basis of designs provided by UK buyer. New designs were being developed for fair exhibition and preparation of catalogue. They were based on observation of catalogues and magazines. They were planning to use revised product designs for better utilization of wood sizes and capacity of transport containers. Also in the process of developing veneer tops for the hardwood product line.

3. Process changes

Firm 41. No changes in production processes due to nature of their manufacturing operation (wholly manual).

Firm 131. No technical changes introduced in the production processes of billiard table.

Firm 231. As an unintended result of a process improvement made the year before to save time, they realized energy consumption savings. The changes were carried out under the direction of the Production Manager, it was done in a couple of days, and no cost data was available. This year they have improved the packaging system which has lowered transport costs. It was also done under the direction of the Production Manager. An unexpected benefit was a substantial increase (about 25%) in the number of beds that can now be shipped in each container. Cost data was not available.

Firm 331. There have been no changes in the production process.

Firm 1191. No changes made in the production process.

Firm 1471. Through small improvements in efficiency they have been able to increase output with the same equipment. Costs have been minimal; the work was all done in-house and the results were very satisfactory and immediately felt. No cost data available.

Firm 471. Four years ago they started to introduce hydraulic and pneumatic assembly fixtures which helped to standardize assembly, and to improve productivity thus reducing costs. Owner saw similar devices in South Africa and had them replicated by a local engineer. The process took about four years to be completed. Costs and benefits data not available.

Firm 1771. They have substituted in the upholstery department polyester fabric, locally available, for imported acrylic (Dralon). There was however no concomitant change in the production process.

Firm 1861. Adaptation to local raw materials resulted in lower costs, but did not seem to have resulted in changes in the production process.

Firm 1921. No changes have been made to the production process. They were contemplating the introduction of energy saving equipment, and to cut overtime to reduce energy consumption.

4. Use of technical support services

Firm 41. Requested assistance from small business office in SEDCO, but they did not respond. Owner is mainly interested in improving his production processes for which he needs a few power tools and to bring electricity to his shop.

Firm 131. Have only used local R&M support services when their machines broke down. Owner unhappy with the service. Argued that they took too long because of lack of spare parts. With respect to a 'wish list' of desirable support services, owner did not express any special needs. He was very upset with the government because of its labour legislation and foreign exchange controls. He would like to be able to fire workers at will, and was particularly annoyed because there was a limit of Z$ 20,000 (about US$ 3,100) for the value of products one can send out of the country as a sample, to be sold in consignment, and fancy billiard tables cost much more than that. He would thus be required to send them as regular exports and to pay the equivalent foreign exchange to the Central Bank, which he of course, did not like to do.

Firm 231. Have used quality control testing and R&M services. The former to test new bed prototypes. It was done at the University of Zimbabwe and they were satisfied with the work done for which they paid a fee. Several local providers took care of their electric motors when they burnt out. They were also satisfied with their services. As to the provision of beneficial technical support services, they ranked them as follows: (1) personnel training, (2) trouble-shooting assistance, (3) quality control, and (4) process improvement.

Firm 331. They have received technical assistance from abroad to improve their production techniques. First, an organization of US retired executives sent a member who stayed for two months. They were not satisfied with the results. This they attributed to the fact that the expert that came was not used to the simpler technologies utilized in this factory, being familiar instead with more automatized processes. The second was a consultant from Holland who, it was claimed, 'in two days did much more than the American in two months', and made good recommendations. However, these were not implemented due to resistance by the Production Manager. In both cases they paid for the services provided. They would rank the services they desire as follows: (1) productivity improvement techniques, (2) personnel training and (3) product redesign.

Firm 1191. Only used R&M service to fix machines as required. Their ranking of desirable support services was: (1) personnel training, (2) R&M, and (3) product redesign.

Firm 1471. Used local R&M shops to fix machines when they could not do it in-house. Their rankings of beneficial support services were as follows: (1) product redesign, (2) trouble-shooting assistance, (3) quality control, (4) productivity improvement techniques, and (5) personnel training.

Firm 471. Used specialized local service to sharpen machine cutters on a continuous basis. Also electrician for R&M of electrical parts and electronic controls in machines. Their rankings of desirable support services were: (1) productivity improvement techniques, (2) labour relations and (3) process improvement.

Firm 1771. Used local R&M services for machines and vehicles. They also took care of air compressors. Satisfied with the service provided. Rankings of desired support services were: (1) productivity improvement techniques, (2) product redesign, (3) manufacture of tools and fixtures, (4) pollution control, and (5) personnel training. (These rankings were obtained averaging the opinions of the three respondents for this firm.)

Firm 1861. They have received assistance from the Norwegian manufacturer of the glue used in veneering. Also received technical assistance for the installation of multiple head lathe, provided by a paid consultant from the United States over a one year period. Regularly used services of local engineering company for R&M. They also arranged for a one week training, provided by the South African office of Siemens, on the operation of the computer-controlled router. They ranked desirable support services as follows: (1) productivity improvement techniques, (2) R&M, (3) energy saving (kilns consume a lot of energy), (4) personnel training, and (5) product redesign.

Firm 1921. Used local R&M services. Hired consulting services of local international accounting and management consulting firm to instal Kawasaki method for production process and quality control. Have arranged with ZIMDEF for an apprenticeship programme involving 10 workers that received a subsidy of 80% of the Grade III salary while they were in training and working for them. Since desirable support services were ranked differently in the pine and hardwood departments, they were averaged: (1) R&M, (2) trouble-shooting assistance, (3) process improvement, (4) productivity improvement techniques, and (5) product redesign.

Based on the rankings provided in Table AIII-6, a numerical index was constructed for the technical support services deemed desirable by the firms visited in the industry. Points were assigned to each rank as follows: 1 = 5 points, 2 = 4 points, 3 = 3 points, 4 = 2 points, and 5 = 1 point.

Table AIII-6 Wood and Furniture. Ranking of desired support services

Firm	*Ranking of support services*				
	1	*2*	*3*	*4*	*5*
41	Process	–	–	–	–
131	n.a.	n.a.	n.a.	n.a.	n.a.
231	Training	Trouble.	Q.C.	Process	–
331	Productiv.	Training	Design	–	–
1191	Training	R&M	Design	–	–
1471	Design	Trouble.	Q.C.	Productiv.	Training
471	Productiv.	Labour rel.	Process	–	–
1771	Productiv.	Design	Tools&fix.	Pollution	Training
1861	Productiv.	R&M	Energy	Training	Design
1921	R&M	Trouble.	Process	Productiv.	Design

Notes:
n.a.: Did not provide any rankings.
–: Did not provide all five rankings.

Table AIII-7 summarizes the results for the 10 wood and furniture plants visited and the top technical support services.

As shown, the responding firms expressed the greatest interest in productivity improvement techniques (six out of 10 firms) and personnel training and product design (also six firms in each case). The highest rankings, among services ranked by at least four firms, were assigned to productivity improvement followed by process improvement, training and product design.

The above results are quite similar to those obtained for the food and beverages, and textile and clothing industries.

5. Training

In the questionnaire a distinction was made between the training needs of operative workers, technical and middle-level personnel, and professional personnel.

Table AIII-7 Wood and Furniture. Overall ranking of support services

Service	*No. of firms ranking it*	*Average score*
Productiv. improvement	6	4
Training	6	3
Product design	6	2.83
Process improvement	4	3.25
Repair and maintenance	3	4.33
Troubleshooting	3	3
Quality control	2	3
Average	**4.28**	**3.34**

According to their training, operative workers can be classified in three groups: (i) those formally trained for a particular job (for example via an apprenticeship programme), (ii) those with experience acquired in a similar job, and (iii) those without formal training and/or experience.

Firm 41. For workers with formal training, without manual experience, one to three months were required to develop the required manual skills. Other workers, with or without similar experience did not require any training. (This is justified by the lack of any powered tools in this operation.) There were no technical or professional personnel.

Firm 131. Workers without formal training and/or similar experience, required one to two years of on-the-job training. No professional or technical personnel were employed by the firm.

Firm 231. Workers with formal training needed four to five years of training. Those with similar experience only required short on-the-job training. For other workers, up to 10 years maybe needed. There was no training for professional personnel. They did provide training for technicians and middle-level technical personnel. They had sent employees to train in lumber milling at the Lumber Institute. At interview time they had 16 trainees from ZIMDEF, most of them in carpentry.

Firm 331. Held the view that only a few days of training were required to learn to operate the plant's equipment. They did not provide training for professional or technical personnel. Complained that the local Polytechnic did not prepare apprentices for the wood and furniture industry, and that they had lost several trained workers to other companies.

Firm 1191. Formally trained workers required about a year on-the-job training; workers with similar experience, four to five years, and other workers more than four years. All are started doing the simpler jobs. They did not provide training for professional personnel. Technicians came up from the ranks and they were provided training. They had plans to provide in-house training for all the plant production processes. Had employed students from Mutare Technical College, and from Belvedere-Harare Institute of Technology. Four years were required for them to be fully qualified, including passing the trade tests.

Firm 1471. Plant manager had not had sufficient time on the job to evaluate the training needs. He believed that for 'general hands' two–three days might be enough. For workers with training or prior similar experience he thought it would depend on each case. No training was provided for professional personnel. General Manager wanted to increase productivity and was sending middle-level personnel to take locally provided short-term courses on supervision.

Firm 471. Workers with a carpentry certificate needed only a couple of weeks. Workers with similar experience required no additional training. Other workers were hired on basis of recommendation from, or kinship with, present workers. There were no educational requirements, and to start, for example in sanding, from a couple of days to a week of training were needed. No training of professional or technical personnel was provided.

Firm 1771. Workers with secondary school and three years of vocational school could be trained in one to two months. Those with similar experience did not require training. For other workers, entering to do unskilled tasks such as sanding and delivery, one month of training was needed. No training for professional or technical personnel was being provided. They were encouraging, but not paying for it, the attendance of secretarial school, after working hours, by one employee.

Firm 1861. They hired school graduates with 'O' level passes plus wood-work knowledge. The training period differed with the job and the individual. For workers with prior experience in similar jobs not much on-the-job training was required. Other workers needed two to three years of training. They had sent one professional for a week course to South Africa. To improve management skills middle-level personnel were sent to a local training institute for management courses.

Firm 1921. Requirements differed between hardwood and pine furniture sections. In hardwood they preferred to hire people with production line experience, especially for cabinet-making. These only require 'induction' training. Workers with formal training, i.e. diploma holders, stay a couple of hours each day with the supervisor for about three months. Skilled occupations in this section were: cabinet-makers, upholsterers, tailors and machinists. In the pine section, skilled occupations were machinists and sprayers. For machinists hired from other companies, training might vary from two to three days up to a month. For sprayers, about two months, while assemblers only demanded a week. When hiring workers for training purposes they preferred graduates with 'O' and 'A' level passes. As an overall summary they held that workers with formal training needed three months' on-the-job training, those with similar work experience, one month, and other workers from one week to two months. They did not provide training for professional or technical personnel.

E. Technological development plans

Only replies containing concrete plans for the future were considered.

Firm 41. No plans, although with electricity and a few power-tools output and productivity could increase several-fold.

Firm 131. No plans.

Firm 231. Would like to expand plant to increase exports of pine furniture to Europe, but using same technology. About Z$ 500,000 in new machinery would be needed. The results would be better-quality products, increased productivity and lower costs.

Firm 331. New plant and had no expansion plans.

Firm 1191. Plan to invest about Z$ 1 million in new machinery. Present stock of equipment was very old. Expected to improve quality, safety, pollution, and competitiveness with new investment. Machines would be more automated and supplier would send training personnel. More technical experts would also be needed.

Firm 1471. Expected to expand and to increase productivity. Did not want very sophisticated equipment. Would also like to hire a designer and to train his personnel to improve their machine knowledge. The total cost would be about Z$ 2 million. Had been in touch with Dutch government programme of technical assistance. Organizational changes required would be more and better supervision, better quality control and the introduction of design function. Quality-wise they expected to be able to meet European standards for exports.

Firm 471. Planned to buy additional machinery and to move to new building for which they already owned the land. The main pieces of equipment to be acquired were: a cutter for Z$ 170,000, a lathe for Z$ 150,000, and a compressor for Z$ 200,000. Land and building cost were estimated at Z$ 4.5 million. Thus the total investment planned was about Z$ 5 million. As a result of this new investment they planned to increase exports, but never beyond 40–50% of total output due to the prevailing differences in prices between domestic and export markets.

Firm 1771. Were planning to expand in local and domestic markets, and to acquire new technology for veneering process, and also to improve their own distribution capacity by acquisition of transport equipment. Total investment expenditure of some Z$ 2.7 million was broken down as follows: Z$ 1.2 million for building and structures, Z$ 400,000 for vehicles, Z$ 500,000 in machinery of various kind, Z$ 120,000 in painting equipment, Z$ 150,000 for automatic lathe, and Z$ 300,000 in other equipment.

Firm 1861. Planned to expand exports to Europe and South Africa. The introduction of a new line of pine furniture for the South African market, to be designed in-house, was envisaged. Investment cost data were not available.

Firm 1921. In pine furniture section they planned to: (i) improve production flow and reduce material handling, (ii) establish satellite plant in Mutare for timber milling, pre-cutting and sorting, (iii) fine-tune the machinery, (iv) instal a dip-painting system, and (v) utilize the sawdust to manufacture their own chip-boards. With respect to the hardwood line they planned to: (i) increase the share of veneer tops using a locally available technology, and (ii) refurbish the machines. Implementation of these plans would lead to: lower costs, better products, increase in exports, greater control of raw material supply and utilization of byproducts. Investment cost data were not available.

F. Technical skills

There was little agreement among respondents about which jobs should be considered 'skilled' and which not. Consequently, in this section we shall skip references to skilled occupations among operative workers and concentrate instead on professional personnel, i.e. employees holding a university, or tertiary degree, and middle-level technical personnel, defined as those holding a certificate or diploma.

However, it should also be noted that a great variety of post-secondary, vocational courses, exist, with different durations and academic requirements, which demands that great care be exercised in handling the information pertaining to middle-level technical personnel.

Firm 41. No professional or technical personnel employed.

Firm 131. Ditto.

Firm 231. One, locally trained, professional accountant and six wood-working certificate holders were employed by this firm.

Firm 331. One, locally trained, unspecified professional, and three diploma holders were part of this firm's personnel. Two of the diplomas were in wood-working, one in upholstery and one in book-keeping.

Firm 1191. No professionals. The General Manager held a diploma in wood-working and there were 14 holders of certificates, also in wood-working.

Firm 1471. No professional personnel. There was one wood-working diploma holder, trained abroad.

Firm 471. No professionals and one diploma holder in carpentry.

Firm 1771. One professional accountant trained in Kenya and Uganda. Two mechanics with certificate, 15 certificate holders in wood-working, and six in business and management.

Firm 1861. No professional personnel. Production Manager, expatriate, held a diploma in wood-working. A mechanic also held a diploma, as well as two others in unspecified specialties. There was one certificate holder in time-and-motion study.

Firm 1921. One expatriate professional held a bachelor of commerce obtained abroad. There was one mechanic certificate holder, and three diploma holders. The diplomas were in: personnel management, accounting and finance, and cost accounting. All obtained locally.

Comparing with Table AII-8 for the textiles and clothing industry, the proportion of professional personnel is lower while that of middle-level technical personnel is substantially higher.

Table AIII-8 Wood and Furniture. Professional and technical personnel

Firm	*Profess.*	*Diploma*	*Certifi.*	*Employ.*	*% Profes.*	*% D + C*
41	0	0	0	4	0.0	0.0
131	0	0	0	14	0.0	0.0
231	1	1	6	197	0.5	3.5
331	1	3	0	195	0.5	1.5
1191	0	1	14	115	0.0	13.0
1471	0	1	0	77	0.0	1.3
471	0	1	0	236	0.0	0.4
1771	1	0	23	107	0.9	21.5
1861	0	4	1	220	0.0	2.3
1921	1	3	1	358	0.3	1.1
Total	**4**	**14**	**45**	**1523**	**0.3**	**3.9**

Notes:
D: Diploma
C: Certificate

Appendix E: Firm Interviews by Sector: Metalworking (ISIC 38)

The eight firms surveyed produced a variety of metal equipment, machinery and metal products.

A. Background information

1. General

As can be seen in Table AIV-1, the firms surveyed differed significantly in size (measured by employment), labour productivity (measured by output per worker), ownership and ethnic origin.

The metal packaging plant had the highest output per worker. Larger plants tended to have higher output per worker.

2. Market orientation

Although Zimbabwe is a country with rich mineral endowment, this had not generated a strong metallurgical industrial base. On the other hand, due to foreign exchange constraints, particularly during the blockade that followed the unilateral declaration of independence, local industrialists and their workers acquired many mechanical and metalworking skills needed to maintain their equipment and to produce spare parts that could not be imported. It should also be noted that, compared to other African countries, Zimbabwe did enjoy a fairly good education system prior and after independence (see Table 2.1 in main text).

Table AIV-1 Metalworking. Basic characteristics of firms interviewed

Firm	*Product*	*Employment*	*Own./ethn.*	*Location*	*Sales/work ($)*
521	Gearboxes	24	Afr./Euro.	Harare	6,400
581	Stain. Steel Vessels	150	State	Harare	12,300
1141	Sew. machin./irons	88	European	Bulawayo	2,100
1641	Pref. house/engineer.	96	European	Harare	11,200
411	Steel/alum.kitch.ware	1353	n.a.	Bulawayo	11,400
491	Metal packaging	528	State/Euro	Harare	29,100
1361	Windmills/dust.coll.	43	European	Bulawayo	8,900
1751	Hammermill	75	European	Harare	14,400

Table AIV-2 Metalworking. Proportion of output exported

Firm	*Exports/$US*	*Sales/$US*	*% Exported*
521	0.0	154,000	0.0
581	80,000	1,850,000	4.3
1141	26,400	185,000	14.2
1641	484,000	1,075,000	45.0
411	5,800,000	15,400,000	38.0
491	300,000	15,400,000	2.0
1361	0.0	383,000	0.0
1751	0.0	1,080,000	0.0
Total	**6,690,400**	**35,527,000**	**19.0**

Metalworking establishments could be classified in two main groups: those manufacturing a specific product, or line of similar products, and those, generally workshops, executing individual projects on request, including the repair and manufacture of machinery and spares, as well as other equipment. In terms of market orientation, the latter metalworking activities were generally oriented to meeting local market needs, while the former were more prone to develop an export capacity.

As can be seen from Table AIV-2, three firms had zero or negligible exports (i.e. less than 1% of turnover), and none reached 50% of its output as the proportion exported. Engineering products in one case, and kitchen-ware in the other, were the products exported by the firms with the highest proportion of exports. The firm producing kitchen-ware claimed to export aluminum pots and pans to South Africa at the same price as they sell them in Zimbabwe. They enjoy an 9% export subsidy and receive tax drawback and foreign exchange benefits. This plant also has a free-bonded zone for imports and production to be exported.

3. Organization

Questions were asked about the existence of professional management in the firm, the organizational structure adopted, and the presence or not of certain formally assigned managerial functions.

As can be seen from Table AIV-3, only two plants (the smallest) did not avail themselves of professional management, and four had several management tiers with functional division of responsibilities.

B. Entrepreneurial history and acquisition of technology

1. Origin of the firm and initial technology

Firm 521. Started as a bus company before independence and moved into repairs and manufacturing of gearbox. Bought some machines as metal scrap and was able to put them back together and into service. Other equipment was bought from Eastern Europe because of price considerations following the advise of owner's friends.

Table AIV-3 Metalworking. Organizational structure

Firm	*Profess. manag.*		*Organizational structure*				
	Yes	*No*	*Direct (1)*	*Super. (2)*	*1 Tier (3)*	*Tiers (4)*	*Other*
521		x		x			
581	x				x		
1141	x				x		
1641	x					x	
411	x					x	
491	x					x	
1361		x		x			
1751	x					x	

Firm 581. Began as a small operation in 1959 by individual entrepreneur and claimed to be presently the largest stainless steel fabricating company in the country, serving the needs of the food, chemical, pharmaceutical, mining and other industries. Founder was still on board of directors but the company had been taken over by the government.

Firm 1141. Family firm. Bought small operation started by original owner-entrepreneur. Kept operating with very old equipment and traditional work-force, even managing to export the simple sewing machines they produced.

Firm 1641. Firm originally established in the 1930s by single foreign owner-entrepreneur and now producing prefabricated housing and various engineering products. A management buy-out took place in 1986 and a reorganization more recently as the firm was experiencing serious difficulties. Additions have been made to the equipment in use.

Firm 411. Originally small metal pressing operation acquired in 1958 by a family group that expanded and capitalized the company. In the 1970s started to manufacture also pipe and plastics, splitting in 1991 into two plants: one producing the kitchen-ware (interviewed) and the other the plastic products. At survey time, the majority political party (ZANU PF) controlled 44% of the shares in the group. The family still managed the operations via a third generation member.

Firm 491. Metal packaging plant, a division of a MNC with majority share-holders in France and part in the United Kingdom. Started in 1958 importing the cans and in 1959 set up a full-blown manufacturing operation occupying about one half of the present premises. A major consolidation and expansion took place in 1982. Technical feasibility study carried out by parent company that also selected the equipment. The parent company goes back to the eighteen century in the United Kingdom.

Firm 1361. Started small, in 1978, with two employees, and grew gradually. He had experience from working in similar metalwork and part of the clientele followed him when he left. Bought machinery as needed, generally, used, locally made machines known to him from experience. Had two partners whom he bought out. Born in Scotland but got his education in Zimbabwe. Went to technical college and had a five-year apprenticeship after high school. Was factory manager of a metalworking company. Has a diversified capability for sheet metal work, including stainless steel and aluminum welding. Presently produced windmills, dust collectors and other sheet metal products, all for the domestic market.

Firm 1751. Member of a local group of companies. Established in 1946 downtown and moved to present premises in 1974. Engaged in manufacturing of hammermills and grain dehullers as well as the provision of automotive engineering services requiring precision grinding, such as camshaft repairs, and special projects on demand.

2. Sources and characteristics of equipment

Since a complete census of all machinery was not feasible, the questionnaire included questions about the three principal pieces of equipment in each factory. Table AIV-4 summarizes the information obtained about origin, age and if the machines were bought new or used.

As can be seen from Table AIV-4, most of the equipment was bought new, and quite a bit locally. Among the main sources of imported equipment were

Table AIV-4 Metalworking. Information about key pieces of equipment

Firm	*Origin*	*Age (years)*	*Bought new or used?*	
			New	*Used*
521	Bulgaria	13	x	
	Bulgaria	13	x	
	Bulgaria	13	x	
581	Zimbabwe	20	x	
	Zimbabwe	25	x	
	UK	15	x	
1141	Zimbabwe	>40	x?	
	Zimbabwe	>30	x?	
1641	UK	15	x	
	UK	25	x	
411	Germany	>20		x
	South Africa	11	x	
491	Germany	12	x	
	UK	35	x	
	USA	35	x	
1361	Zimbabwe	5	x	
	Sweden	15		x
	USA	20		x
1751	Imported?	5	x	
	Imported?	10	x	
	Imported?	10	x	

the United Kingdom, Germany and the USA. While a majority of the main pieces of equipment seemed to be less than 20 years old, two factories had principal equipment which was 30 or more years old.

3. Use of foreign technology

The questionnaire inquired about the existence or not of foreign licences or technical assistance agreements, and whether the responding firms had received help for commissioning the plant or installing the equipment.

Firm 1641. Had manufacturing licences from Australia and the USA, as well as a technical assistance agreement. Paid 5% of sales as royalty. Their contract contained restrictions on imports (to guarantee quality), exports (only allowed to African countries), and provisions for the cross-transfer of know-how. They also received help to instal new equipment. Suppliers came to inspect what had been done, and also to assist in establishing the new processes and in training plant personnel. Top managers also travelled abroad to learn the operation of the new equipment.

Firm 411. Paid royalties (5% of sales) for enamelling process and used to have a licence to manufacture cookers. Got European Community technical assistance for some time. Also got help from Spanish supplier for installation of new polishing equipment. Only the local expenses of these experts were paid by the firm.

Firm 491. UK headquarters appointed the Technical Manager who is the advisor and trainer in technical matters. They usually stay for two years, but have asked for an extension. They received help from supplier for commissioning new equipment, and the Technical Manager was in close contact with the project engineer in the UK.

Firm 1361. Have applied for a licence for the manufacture of windmills without generator from Norway, but it has not been finalized yet.

As can be seen in Table AIV-5 only two of the firms interviewed had licence agreements, while three had technical assistance agreements and had also received external help for the installation of new equipment.

Table AIV-5 Metalworking. External technical assistance

Firm	*License*		*Technical assistance*		*Outside help*	
	Yes	*No*	*Yes*	*No*	*Yes*	*No*
521		x		x		x
581		x		x		x
1141		x		x		x
1641	x		x		x	
411	x		x		x	
491		x	x		x	
1361		x		x		x
1751		x		x		x

C. Operating capabilities

1. General

Firm 521. Spacious, well built and with good lighting, airy, clean and well-organized plant. Equipment of mixed ages, some 15 and other 40 years old. Worked one shift and claimed a 75% rate of capacity utilization. No idle machines or any other unutilized equipment on view.

Firm 581. Good building, safe floors, clean. Machinery quite old, most was 20–30 years old. Worked one shift, and stated they were operating at only 40% of their capacity. Some machines were idle for lack of orders and frequent breakdowns.

Firm 1141. Very old premises in state of disrepair. Plant looked messy and assembly line crowded. Equipment was quite old, most of it more than 40 years old. Plant operated one shift at 50% capacity. Assembly line was idle awaiting sewing machines arms and heads. Daily output varied greatly depending on breakdowns.

Firm 1641. Nice plant with excellent ventilation and lighting. Machines were 15–20 years old, but equipment in machine shop was quite new (5 years old). Operated one shift at about 50% capacity. Of seven production lines only three were in operation and there were idle machines also in the machine shop.

Firm 411. Spacious and well-organized plant. Machine parts were colour-coded. Had their own rail-track and a conveyor for enamelling process. Missing protective guards in presses. Equipment was quite old – 40–50 years on average. Worked two nine-hour shifts. Average capacity utilization was estimated at 85% (100% during the day shift and 70% during the night shift). Little or no idle equipment were apparent.

Firm 491. Well-lighted and clean premises. Production lines were well organized. Recently installed KPS which changed the plant layout into a cells system. Workers of the different cells were recognized by the colours of their overalls. Equipment was 20–25 years old, with one new line that was not in operation due to lack of demand. Most of the plant operated one shift but the coating and printing department were operating three shifts. Capacity utilization varied greatly per section, and was estimated to be around 60% on average.

Firm 1361. Spacious, well-lighted plant, but painting area had no air-extractor and the disposal of sand-blasting residues was poor. Stores area was a bit messy; inventories of parts in use were mixed with old stuff and scrap. Machines were on average more than 15 years old. Plant operated one shift with some overtime during the weekend. Capacity utilization was estimated at about 80%. Many machines were not in use, but it was attributed to the nature of the work, plus some workers were out installing a dust collector of their fabrication.

Firm 1751. Dirty and disorganized plant with machines scattered all over the place. Well-organized stock room using computers for inventory control. Painting was done outside without the benefit of a protective booth or any extraction system. Some equipment was less than 20 years old, but most of it was quite old. Operated one shift, and the overall average rate of capacity utilization was estimated at 65%. Idle machines were evident.

2. Troubleshooting

Firm 521. Oil pump broke down but was repaired in house.

Firm 581. Have experienced equipment breakdowns and production bottlenecks. The equipment was fixed in part in-house (they have machine shop), and also by supplier. The bottlenecks slowed down production, but they have not solved the problem yet. Had notified IDC.

Firm 1141. Breakdowns, bottlenecks, quality, and pollution problems. Equipment breakdowns occurred daily and were due to the very old age of the machines. The plant manager saw his role as that of a fire-fighter. He could not plan maintenance or take care of other tasks besides coping with machine breakdowns. One of his main problems was the re-deployment of personnel after a breakdown. Workers were on average some 55 years old, quite set in their ways and many could only do a few tasks, so when there was a breakdown, the plant manager sometimes had to shift five or six other workers to keep one of them occupied. Machines were repaired by in-house personnel and also with outside local help. Iron poured manually lead to high reject rates; about 10% of the castings had to be remade. Other quality problems arose because their old machines lacked accuracy. Also because their old age labour force lacked in strength and skills. Have also experienced assembly problems because sewing-machine parts were bought from various international sources (China, India, Taiwan), and their measures varied slightly but enough to cause operating difficulties. Now before buying new parts they were testing them. Bottlenecks were caused in part by the constant breakdowns. Also foundry could produce a lot more but they did not have space to do any additional casting. These problems had not been solved yet. They were dumping sand casting residuals behind the plant in their own land. Although not causing an immediate health problem, the dumped materials would have to be removed before any new construction could take place on that site.

Firm 1641. Had experienced breakdowns, quality problems, bottlenecks and minor safety problems. The equipment breakdowns were minor and they repaired them in-house and with outside local help. Quality of product was questioned by customer who returned it; they reworked the item in the engineering section. There was an imbalance in the metal cutting section (guillotines) which led to a slow-down in the work. Problem of apportioning the time of these machines to each section had not been solved yet. They had minor accidents because the corrugated iron cuts through the gloves in use by the workers.

Firm 411. They experienced breakdowns, as well as safety and pollution control problems. Attributed the continuous breakdowns to human error by tool setters and foresaw the need for training and retraining. Safety problem was also continuous and had not been satisfactorily solved. Every year workers lose fingers, specially in the continuous production lines. Accidents were more frequent during the hot spell in October. Interviewers noted the lack of safety guards in presses (see above C.1). Had problems with the disposal of acid and other effluents from pickling. Supplier advised them about how to solve the problem. Also had difficulties with smoke emanating from enamelling ovens. It damaged neighbours' roofs and they got fined for this.

Firm 491. Had problems with electrical control system and with welding machine in open top line. For the first problem they got expert technical help from South Africa. The second was solved by the technical manager who was an expert in welding. Formal training was also introduced.

Firm 1361. Had difficulties with air plasma cutting machine that was down for two weeks. It had been manufactured in Italy for UK supplier. Received wiring drawings for the machine from UK but they were not correct. Their own people then fixed the problem.

Firm 1751. Experienced difficulties with subcontractors that did inferior jobs which they had to redo. Also, a large lathe broke down and it took two months to get it fixed. A supplier helped them find the required gears, which they could not manufacture, and they installed them.

3. Quality control

Final 521. Final product, production processes, materials purchases and materials and components in process were all subject to visual, practical inspection by owner-manager. Did not use any norms or standards, and there was no routine for sampling inspection. Claimed that their gears were superior to those made by other local producers because they were the first to manufacture them in the country. Held their quality to be comparable to that of imports. Customer complaints were dealt directly with them. There was no system for statistical quality control (SQC). No data was available on internal rejects.

Firm 581. Materials, processes, components and final product were subject to visual, practical inspection. However, for pressurized tanks they had to meet SAZ and BSI 700 (British) standards. These were required by law and they claimed to have adopted such norms 35 years ago. Sampling inspection was only done for sanitary ware. Surprisingly, they admitted having lower quality than local competitors, and much lower than imports. Internal reject rates were between 10% and 20% three years ago, and around 3% last year. The improvement was attributed to the hiring of a new production manager three years ago. Customer complaints were handled through the sales force. There was no SQC system.

Firm 1141. The final product, raw materials and component purchases, and production processes, were subject to visual, practical inspection. No standards were in use. They carried out sampling inspection during the production process. Had no local competition for sewing machines and claimed to have quality comparable to imports such as Singer. Had about 10% rate of internal rejects which were reworked in the plant. Claimed to have no rejects after the machines leave the plant. Complaints were handled by direct contact with customers. Recent complaints about the sewing machines were about stitch length and due to the control mechanism not working well. No system of SQC in place.

Firm 1641. Materials and components in process, as well as the final product, were subject to visual, practical, inspection. There was no control of processes or raw material purchases. Ladders were subject to SAZ and British (BSI 700) standards. Engineering products were also subject to standards. Their application was voluntary. They used sampling inspection but had no SQC system. Felt that their quality was comparable to that of local and import competitors. From 5% to 10% of the engineering division's output was returned and fixed. Claimed to have no final product rejections. Solved complaints by direct contact between management and customers.

Firm 411. Final product, processes, raw materials and component purchases, as well as materials and components in process, were generally submitted to visual, practical, inspection. Additionally, propane containers, the enamelling process, and raw material purchases were subject to various laboratory tests. They applied appropriately modified SAZ, South African, and British standards. Firm's technical representatives participated in the SAZ technical committee for electrical appliances. The standards in use were compulsory for gas cylinders, and voluntary for other products. Sampling inspection was applied to propane containers and one of every 200 cylinders was tested for destruction. Leakages, expansion, and enamelling of other products were also tested by sampling. They believed that their quality was higher than that of local competitors, and comparable to that of imports from South Africa and China. Claimed that their market share of aluminum ware was increasing. Internal rejects were sold as 'seconds'. They estimated internal rejects rates of about 7% for aluminum ware, and greater than 4% for enamelled products. Customer complaints were handled through the sales force. There was no system of SQC.

Firm 491. The final product was subject to visual, practical, inspection as well as to laboratory tests. Processes were tested on line with appropriate equipment. Raw materials and component purchases were also inspected in a visual, practical, manner. Materials and components in process were subject to laboratory tests. They used the standards of their British headquarters. Samples for inspection were taken during the production process every 15–20 minutes. Tests included: length, width, welding, seaming, lacquer application, etc. Had no local competition and claimed to have higher quality than imports. Scrap loss was less than 1%, and rejects requiring reworking were between 2% and 3%. Complaints were handled by the sales force, and also by direct contact with customers. There was no system of SQC, but they had a line audit with quality control checks as part of the KPS they had adopted.

Firm 1361. Final product, processes, and raw materials and component purchases were subject to visual, practical, inspection. Used no standards. There were no sampling inspection routines. Claimed to be unaware of how their quality compared to that of competing producers. For products such as dust collectors, claimed to have no competition. For windmills they offered a six years warranty. Reject rate was estimated at 1%. Customer complaints were directly handled by the general manager. There was no SQC system.

Firm 1751. Final product and processes were subject to visual, practical, inspection, plus various measures and tests. Some of the tests were carried out in house, others, such as hardness, were sent out to SAZ. They used national norms and were members of the Engine Assembly Association which controlled motor quality standards of voluntary application. Had no sampling inspection routines. Claimed to have higher quality than local competitors. Last year had a 5% reject rate. Had improved by upgrading work-force skills. Believed that many problems were due to the way the equipment was handled upon delivery. Customer complaints were dealt with by the sales force, but the works manager and the sales manager also met to discuss the problem. There was no SQC system.

4. Repair and maintenance (R&M)

Firm 521. The owner-plant manager was in charge of R&M. Claimed to check and service the machines daily. Breakdowns were infrequent and the problems were fixed without outside help. Claimed to have a complete R&M capability in-house.

Firm 581. Had a machine shop and its foreman was in charge of R&M. Claimed to provide complete R&M services in house but also required outside specialized repair services (Machinery Tools Association). They claimed to service the machines monthly. Frequency of parts replacement varied. Breakdowns were frequent (two to three times per month) but generally it was the same machine (guillotine).

Firm 1141. The factory manager was in charge of R&M, and keeping the equipment running was his main activity. They claimed to have the capacity to strip most machines to replace parts. Machines were continuously checked and serviced daily. The high frequency of breakdowns (daily) was due to the very old age of the equipment. They completed most repairs in house but sent electric motors to be fixed outside.

Firm 1641. A maintenance supervisor and his team serviced the machines and kept them going. Every three months they had oiling, etc. A shutdown and overhaul period lasting two to three weeks takes place at each year's end. For the overhaul they brought outside experts. Breakdowns were not frequent (about every six months).

Firm 411. The Works Director was in charge of R&M. Plant included a large machine workshop and they claimed to have the capability to provide complete R&M service in house. Machines were only checked when they broke down, but they had a schedule for oiling, daily, before the start of each shift. They also knew which spare parts were needed and kept a good stock of them. Used outside specialized services for electronic controls. Breakdowns were quite frequent.

Firm 491. Maintenance function was decentralized to the three production sections. Oiling was done daily but parts were replaced when order fulfilling process permitted it. Expected to instal a preventive maintenance system with assistance from abroad. Breakdowns were frequent but they normally could fix them quickly by themselves. Outside help was required to manufacture spares, otherwise they could strip most machines and replace parts as required. A specialized company was hired to provide maintenance services for their compressors.

Firm 1361. Two maintenance men reported to the plant manager. Oiling was done weekly, but no major overhauls were planned. Breakdowns were infrequent. Had the capacity to strip most machines and replace parts as needed. Outside help was required for electrical and other specialized maintenance work.

Firm 1751. There was no specific section devoted to R&M but used for that purpose the personnel and machines in the factory shop. There was a shutdown period every year and each operator was charged with servicing his machines. Had the capacity to strip most machines and to replace parts as needed. For repairs of large machines they hired local engineering company.

Breakdowns were quite frequent because a large proportion of the equipment was old.

5. Industrial engineering

Firm 521. No organized effort to improve productivity. There were no time-and-motion studies. No advance scheduling of production. Monthly management meetings decided on stock levels. No computers were in use for inventory control. Layout had been modified recently as a result of plant expansion.

Firm 581. No organized effort to improve productivity, and there were no time-and-motion studies. No advance scheduling of production. Had a Cardex system for inventory control of raw materials, work in progress and the final product. Layout was changed on occasion of plant expansion or replacement of equipment.

Firm 1141. Claimed to have a technician and a practically-trained technician assigned to the task of continuous productivity improvement. No time-and-motion studies were carried out. Set production standards based on experience. They did not know how their output per worker compared with that of local competitors, but believed it was lower than that of foreign ones. Forward planning of production at least one week in advance to meet production or stock targets. Had a Cardex system for inventory control. Plant layout was not being continuously assessed.

Firm 1641. One technician was permanently assigned to the task of productivity improvement. Claimed to have had time-and-motion studies done by a retired US engineer, but to use production standards worked out on other basis a few years back and still in use. Believed to be 40% more productive than small local competitors, on par with large local ones, and were unaware how they would compare with foreign producers. Claimed to have advance scheduling of production based on delivery dates established for orders received. Had Cardex system for stock control of raw materials and final product. Claimed to change layout frequently in order to increase productivity. However, they also stated that they did so when new lines and products were put into production.

Firm 411. No organized, continuous effort to improve plant productivity. No time-and-motion studies. Production standards were based on past experience. Annual salary increment was partly based on the extent to which the annual target was met. The plant manager and the director of operations were in charge of standard-setting. Were unaware of how their productivity compared with that of local or foreign competitors. Planned production at least one week ahead. The purchasing programme was prepared monthly, and for export sales they programmed at least three months in advance. Stock of work in progress was kept with a Cardex system, while inventory control of raw materials and final product was computerized. Every six months they had physical stock-taking. Plant layout was examined on occasion of plant expansion or replacement of equipment. Previous year they expanded the aluminum-ware department increasing its production substantially, and did the layout work by themselves.

Firm 491. Had introduced KPS by a task force and with help of outside consultants. Production manager set production standards based on experience and the speed of the various machines. Believed that their productivity was

75% below that of UK producers. Advance scheduling of production was done by the sales force who got the orders and lead time depended on item. Production with long-term targets could be scheduled three months in advance. In other cases it was down to four–six weeks, and there was a weekly production plan. Ordering of some imported materials, such as plate that comes from South Africa, must be done well in advance because it may take several months to get them. Had Cardex and computer system for inventory and production control. Had 12–15 PCs, of which three were used for inventory control. Layout was changed completely as a result of implementing KPS. Otherwise, plant layout was only examined in case of expansions or replacement of equipment.

Firm 1361. No organized effort to continuously improve plant productivity. No time-and-motion studies. Standards were set by experience by the plant manager. Workers were paid per hour, and there was a 'worker of the month' incentive scheme for the skilled and semi-skilled staff. Were unaware how their productivity compared with that of local or foreign competitors. They planned for the orders at hand, otherwise they produced for stock. Kept inventory control with Cardex system and job cards for raw materials and work in progress respectively. Layout was examined only in case of plant expansion or equipment replacement. Last time done five years back when they bought new press.

Firm 1751. No organized effort to improve productivity. No time-and-motion studies. Standards of production were set by section heads and supervisors based on experience. Did not know how their productivity compared with that of other producers, both local and foreign. According to items, they planned production one month or one week in advance. Had computerized stock controls for raw materials and final product. Used Tetra 2000 computer software package provided by headquarters. Computer applications included, besides inventory control, creditors and debtors, payroll, job costing and general accounting ledger. PCs were also used by financial manager and works manager. Layout was normally examined when there was a plant expansion or replacement of machinery. Recently they had rearranged machines in the general section of the factory.

6. Industrial safety and pollution control

Firm 521. Responsibilities for these functions had not been assigned, and there were no written rules either for safety or for pollution control. No periodic drills for fire evacuation. Believed that their safety record and procedures were comparable to those of other firms in the industry.

Firm 581. Had a nurse in charge of safety; there was also a safety committee. Pollution control was handled in the same way. Had written rules for safety but not for environmental control. There were no periodic drills for fire evacuation. Believed that their safety record and procedures were at a higher level than those of other firms in the same industry. Did not know how they compared in pollution control matters.

Firm 1141. Responsibility for safety had been delegated to the plant foremen. It had not been assigned in the case of pollution control. There were no written rules either for safety or for environmental control. No fire evacuation drills were carried out. Believed that their safety record and procedures were about the same as those of other firms in the industry. In pollution control they felt that they were at the same or lower level than other firms in the industry.

Firm 1641. There was a safety supervisor to insure that workers put on their protective clothing. Believed that they did not generate pollution. Had signs and posters about safety on plant's walls, but no written rules. No written rules for environmental control. They claimed to have trained fire-fighters, but respondent was unaware about the frequency of fire drills. Believed that their safety record and procedures were at a higher level than those of other firms in the industry. Believed they did not generate pollution. One of their workers had been affected by painting fumes, but they blamed the worker for not taking the required precautions.

Firm 411. Responsibility for safety and pollution control rested with the Works Director. There also was a Safety Committee that met monthly and included workers representation. There were written safety rules for crane operators. Claimed to have pollution control rules in preparation. The safety committee had instructed personnel on the use of the fire extinguisher available in the plant. They had no sprinklers. Believed that their safety record and procedures were at about the same level as those of other firms in the industry, but at a lower level on pollution control. They have had problems with neighbours because of emissions from enamelling ovens.

Firm 491. Safety responsibility was assigned to the Factory Manager. There was a risk officer and also a safety committee, which was part of the workers and management committee. Responsibility for pollution control had not been assigned. They were expecting an expert from UK to conduct an audit of the plant in matters related to pollution resulting from painting and coating. They had no written rules, either for safety or for environmental control. They had a designated fire fighting team, and the supplier of fire extinguisher came twice a year to conduct fire drills. Believed that record and procedures on both safety and pollution control were at about the same level as those of other firms in the industry.

Firm 1361. Neither the responsibility for safety nor for pollution control had been assigned, and there were no written rules for either one. Had no fire drills but had trained personnel on the use of fire extinguisher. Believed that their record and procedures were at about the same level as those of other firms in the industry for both safety and pollution control. Interviewers noted problems with the extraction system in the painting section, and also with the disposal of sand-blasting residues.

Firm 1751. The Works Manager was responsible for safety. They also had three employees trained in first aid, and there was a plant safety committee. Responsibility for pollution control had not been assigned. There were no written rules for safety or environmental control. Had protective equipment for sprayers. A worker had recently cut a finger. They had fire extinguishers and the Works Manager was supposed to conduct fire drills. Believed that their safety record and procedures were about as good as those of other firms in the industry, and were unaware how they compared in pollution control.

D. Technological functions and organization

1. General

Firm 521. No codification of technical information. There were no files and no technical offices.

Firm 581. Had written manuals with product specifications. No technical offices. Technical staff was also in charge of marketing.

Firm 1141. No documentation existed for old product line. Planned to document technology for new products. Claimed to have pilot or experimental line with some staff assigned to it.

Firm 1641. Claim to have codified engineering technology to keep quality standards. A total of three staff were assigned to technical functions. One and one-half to product engineering and one and one-half to R&D and experimentation.

Firm 411. Technical documentation included production schedules given to each foreman as part of his job description, drawings for tool and die making, and specifications and other testing requirements used in the quality control laboratory. Three staff in the quality control laboratory. Pilot plant was run by the quality control laboratory under the supervision of the Enamelling Manager. The tool and die-making office included a practical engineer and a self-trained draftsman using CAD machine.

Firm 491. Technical documentation included: product drawings and specifications, tool and die drawings, and operating instructions. Technical offices and their staff were: project engineering, with chief plus four employees, tool and die making with 10 workers, drawing with three staff, library and technical standards with one staff, and food and packaging technology with one staff. They also offered food laboratory service to their customers (food processors).

Firm 1361. Have detailed blueprints for every product. General manager prepared the drawings that were given to the plant manager and the workers. Additional instructions were provided verbally.

Firm 1751. They had specifications for spares in the automotive section and workers were told what to do verbally. Special projects required drawings which were made by the works manager.

2. Product design and product change

Firm 521. They copied similar imported products. Had not introduced modifications or new products in the last five years.

Firm 581. Used self-generated domestic product specifications and had made some not very significant modifications to meet market demands. Had not introduced any new product in the last five years.

Firm 1141. Copied similar imported products with some minor changes. Were contemplating introducing a new product to the market that had been developed in house.

Firm 1641. The sources of their products design were: foreign specification, self-generated according to domestic specification, and copies of similar imported products. On the latter case they tried to avoid existing patents. Modifications introduced were not very significant and made to suit local tastes and to make possible the use of local materials. They had recently introduced new evaporating air conditioners, or coolers, with the assistance of Australian expertise for product development and other outside help for the testing of prototypes.

Firm 411. They copied similar imported products and also self-generated designs according to domestic specifications. They regularly attended international trade fairs in their line of business and made their own designs and prototypes. They had introduced substantial modifications on an electric stove originally produced under foreign license. They changed from hand fabrication to machine pressing using their existing equipment. They had also made the controls simpler and the construction more robust to adapt to local conditions. Took into account features in European stoves. In the last year they had introduced a new line of garden furniture developed in house.

Firm 491. Changes were only in the design of labels or printing on can. For these, customers provided the specifications and art-work and they produced samples in house. In the last five years they had introduced a shoe-polish can with a hand lever for opening. They copied the design. The hand levers were imported and they punched them onto the can.

Firm 1361. A dust collector was copied from a South African model, but they introduced substantial modifications to improve efficiency and adopted better filter materials. For the boats, they originally manufactured a couple and showed them to fishermen who liked the design and thereafter ordered the ships from them. They were considering the manufacture of a new type of windmill of Norwegian origin.

Firm 1751. Introduced last year a higher capacity de-hulling machine with almost double power. The design was self-generated and adapted to domestic specifications.

3. Process changes

Firm 521. Lacking a grinding machine, they adapted the process of gear cutting to use a lathe for grinding.

Firm 581. No process modifications were carried out.

Firm 1141. Ditto.

Firm 1641. Claimed to have introduced in the last five years three process technical changes: substitution of cheaper local fibre glass for previously imported Australian product, use of a power factor correction board to save energy, and setting up a machine shop to manufacture in house the required tools and dies, while also helping with maintenance and product development. They only needed outside assistance for the raw material change. In all cases they obtained cost savings as expected.

Firm 411. Have planned a change in pressing operations, now done manually, to reduce time and increase productivity. Machine and tools expected to arrive from supplier located with European Community assistance.

Firm 491. With aid of outside consultants during 1992–93 introduced a Japanese production management system in the plant. Claimed that the installation was done in six weeks, and had a one year contract with the consultants. Savings were mostly from reduced stocks and increased operational efficiency. Claimed to have already saved $Z 400,000. Personnel was motivated to attain production targets by yearly bonus of up to 10% of gross yearly salary.

Firm 1361. Respondent not aware of any changes in production process.

Firm 1751. There have been no changes in the production process.

4. Use of technical support services

Firm 521. Have only used locally imparted training seminars for workers but were not happy with the results claiming that the level was too low. Having managed the factory for 13 years without problems they did not see the need for technical support services.

Firm 581. Had received quality control and technical aid from abroad. The Standards Association of Zimbabwe (SAZ) provided testing and certification services, for a fee, as required by the vessels they manufacture. A Dutch consultant carried out a feasibility study to improve the manufacturing process and plant layout. It was free of charge but was not implemented, although they could clearly benefit from improving the plant layout. They ranked desired technical support services as follows: (1) new machinery (it is really financing what they believe they need), (2) productivity improvement techniques, (3) R&M, (4) QC, (5) trouble-shooting assistance.

Firm 1141. Used local laboratory QC services to test castings for a fee. Also used local R&M services to fix motors that burn out. Their parent company provided occasional technical assistance when problems were identified. In terms of support services they really only desired to have better trained personnel.

Firm 1641. Had received QC, technical assistance and personnel training services. SAZ visited the plant every six months and tested their products to verify compliance with BSI 700 norms. For new air conditioner they received technical assistance from Australian firm to learn how to operate new machinery. Top management training courses were imparted by a US firm when the new line of production was installed. All the above services were for a fee. They expressed interest in various technical support services: calibration of control instruments, troubleshooting assistance, productivity improvement techniques, tools, dies and fixtures manufacture, product redesign, process improvement and pilot plant experimentation. However, they claimed that all were equally necessary and could not be ranked.

Firm 411. Have used QC, training and technical assistance services. Had used the services of National Railways of Zimbabwe laboratories in Bulawayo for occasional testing of their products. Received computer training services from Zimbabwe Institute of Management and another private provider. Technical assistance, free of charge, was provided by EEC via CDI that provides such help to firms in LDCs in kind or technical services. They were helped to locate suppliers and provided equipment assistance, etc. On one occasion they got shipment of drawing steel not of the required quality and had to anneal it prior to use it because it could not be sent back. Two managers offered different rankings of desired technical support services which were averaged to provide the following result: (1) tools, dies or fixtures manufacture, (2) productivity improvement techniques, (3) product redesign, (4) personnel training, and (5) trouble-shooting assistance.

Firm 491. Used services of SAZ for compression testing of drums; continuously used local R&M services and have used several local companies for calibration of their instruments. Local company imparted computer training courses for their personnel. They would like to have access to yet unavailable local experts for assistance with troubleshooting, but no other support services were ranked.

Firm 1361. Local company had occasionally conducted pressure testing of their vessels used for chemical products. A local company had provided training

Table AIV-6 Metalworking. Ranking of desired support services

Firm	*Ranking of Support Services*				
	1	*2*	*3*	*4*	*5*
521	n.a.	n.a.	n.a.	n.a.	n.a.
581	New equip.	Productiv.	R&M	QC	Trouble.
1141	n.a.	n.a.	n.a.	n.a.	n.a.
1641	n.a.	n.a.	n.a.	n.a	n.a.
411	Tool, dies	Productiv.	Prod. desig.	Training	Trouble.
491	Trouble.	–	–	–	–
1361	Safety	Energy sav.	Productiv.	Pollution	–
1751	QC	Training	Process	R&M	Safety

Notes:
n.a.: Did not provide any rankings.
–: Did not provide all five rankings.

for their welders. A local engineering company provided R&M work and an electrician was under contract for all their electrical work. The general manager ranked the desired technical support services as follows: (1) industrial safety, (2) energy saving, (3) productivity improvement techniques, and (4) pollution control.

Firm 1751. Had received R&M services from local provider to repair big lathe. Also testing for QC and special hardening from local company. Headquarters office (they belong to a local holding group), provided management and computer training courses and seminars. Their rankings for desirable technical support services were as follows: (1) quality control, (2) personnel training, (3) process improvement, (4) R&M, and (5) industrial safety.

Based on the rankings of Table AIV-6 a numerical index was constructed for the technical support services considered most desirable by the eight firms visited in the metalworking industry. Points were assigned to each rank as follows: 1 = 5, 2 = 4, 3 = 3, 4 = 2, and 5 = 1.

Table AIV-7 summarizes the results for the firms in the industry and the top technical support services.

Table AIV-7 Metalworking. Summary of rankings of support services

Service	*No. of firms ranking it*	*Average score*
Productiv. improvement	3	3.67
Troubleshooting	3	2.33
Training	2	3.00
Safety	2	3.00
R&M	2	2.50
Quality control	2	2.33
Average	**2.33**	**2.80**

As indicated above and in Table AIV-6, relatively few metalworking firms expressed an interest in getting technical support services. About 40% provided no rankings. This could be an indication of a professional sense of greater confidence in dealing with technical matters than in other industries.

In terms of their expressed preferences, only productivity improvement techniques and trouble-shooting assistance were ranked, at least, by three out of the eight firms, and only the former got a relatively high average ranking.

5. Training

In the questionnaire a distinction was made between the training needs of different types of labour, i.e.: operative workers, technical and middle-level personnel, and professional employees. With respect to operative workers they were classified in three groups according to their training: (1) those formally trained for a particular job (for example via an apprenticeship programme), (2) those with experience acquired in a similar job, and (3) those without formal training and/or experience.

Firm 521. They kept two apprentices at all times. No training programmes for technical or professional personnel. Workers with only formal training (say for about three years) required 18 months in plant training. For other workers, with some experience, shorter periods of time were required. The shortest period, considered exceptional, would be six months.

Firm 581. There were no training programmes for technical or professional personnel. For workers with experience in similar jobs, little or no training was required. For those without such experience, training needs would vary according to machine but they envisaged between three and four years of on-the-job training.

Firm 1141. There were no training programmes for technical or professional personnel. Inexperienced and untrained workers started as helping hands and required about six weeks to become useful. The best could then operate machines. In an exceptional case one such worker become a qualified welder within two years. Formally trained workers require little or no training, and for those with prior experience in similar jobs, it would depend on the individual.

Firm 1641. They had no training programmes for technical or professional personnel. Formally trained workers needed about six months of training. Those with similar job experience did not need in plant training. For other workers, i.e. with neither formal training nor job experience, four to five years' in-house training were deemed necessary.

Firm 411. Professional personnel received locally imparted management and computer training courses. Similarly for technicians. They attended locally imparted computer courses, and for enamelling training had been sent to South Africa. Formally instructed or experienced workers did not require additional training. For other workers they did not insist on 'O' level requirements because in their experience educated people were not well-suited to routine manual work. They had a set training format which lasted one week, including the induction period.

Firm 491. Professional personnel were provided additional formal training as well as exposure to factory training abroad. Staff had been sent to the United Kingdom and South Africa for such training. Technicians and other middle-level technical personnel were also provided with such training when new lines and new machines were commissioned. Factory in Kenya provided such training recently. They only hired formally educated workers with three 'O' level pass examinations (in English, Mathematics and Science). Such workers normally required one year in plant training, although some would only demand days. Workers with similar job experience did not need additional training.

Firm 1361. There were no training provisions for professional or technical personnel. Formally trained workers (journeymen) did not require in plant training. Workers with similar job experience did not need much training time. Unskilled workers started as 'casual' workers and it generally took a couple of years for them to reach the semi-skilled level. Believed that unskilled workers did not need 'O' level examinations; the ability to read and write was sufficient for them. They provided outside training for semi-skilled workers, in welding, for example.

Firm 1751. There were no training programmes for professional personnel. They provided in-house apprenticeship training; had five in general line and three in automotive product line. Hired workers via a private employment agency and their plant training would depend on their level. No detailed information provided.

E. Technological development plans

While some firms expressed desires about impending technical changes, only those replies containing concrete information about future plans were considered below.

Firm 521. Had been contemplating the acquisition of a grinding machine to improve their process of gear manufacturing. The cost was about $US 12,000. This would permit them to expand output while otherwise following similar production methods. Said to be deterred by high interest rate, and were waiting for financial conditions to improve. Present personnel could use the new machine and they did not envisage the need for a long or complicated learning process. Expected that with the new machine their gears would have somewhat lower costs and somewhat higher quality.

Firm 581. They would like to expand with help of a joint venture and seemed to have been looking for partners in South Africa and Denmark. Plant had apparently surplus labour and technical assistance study recommended layoffs. Plant was owned by IDC, a government entity, and that constituted its main constraint. Thus, without privatization present aims were likely to remain more in the nature of a wish list than concrete plans for the future.

Firm 1141. No plans for technological development.

Firm 1641. New production manager envisaged the need to expand and modernize in order to cut costs and survive in the new, more competitive, environment. Foresaw total investment of approximately US$ 750,000, broken down as follows: US$ 230,000 for new machinery, US$ 385,000 for new building, US$ 115,000 for R&D costs, US$ 12,000 for computers to be used in design. After the new investment they expected that their costs would be 15–20% lower and there would also be some improvement in quality. The extent to which these plans had firm financial support was not ascertained as the firm seemed to have recently been experiencing difficult times and operating substantially below capacity.

Firm 411. Were contemplating the introduction of a new line of products with internal teflon coating and protected exterior, which had superior characteristics. CDI had put them in touch with three potential suppliers of raw materials and equipment. The investment required could vary between US$ 700,000 and US$ 1.5 million according to the supplier selected. They planned to let the market decide because they were expecting to start by importing pre-coated metal circles for pressing and manufacturing. No one had a similar product in the region and they claimed that coated aluminium ware was very popular. They had already allocated space within the plant for the new line.

Firm 491. Although they did not have plans for expansion or technological change, they did have capital expenditures plans which included US$ 120,000 for computers and US$ 90,000 for a grinding machine in the tool and die making shop. The rest, for a total of US$ 1.2 million, would be for production machinery. These new investments will permit them to have more flexibility in the production lines, to standardize templates, and to improve welding control. It was not firm, but could receive external assistance from HQ and perhaps from suppliers.

Firm 1361. Would like to have new, upgraded, equipment for some processes, but had no firm programmes for that purpose.

Firm 1751. Interested in expanding the service of bigger automotive units, in particular, engines of transporters' trucks. Would need bigger machinery for that purpose. Expected help from DANIDA (Denmark). Total investment required was estimated at about US$ 1.4 million.

F. Technical skills

Substantial discrepancies exist among respondents with respect to which operative jobs should be considered 'skilled', and which not. Thus in this section we shall concentrate on professional personnel, i.e. employees holding degrees from a university or other tertiary-level institution, and middle-level technical personnel, defined as those holding a certificate or diploma.

It should be noted, however, that a great variety of post-secondary, vocational courses, exist, having different durations and academic requirements. This demands that great care be exercised in handling the information pertaining to middle level technical personnel.

Firm 521. There was one engineer trained abroad. Also, three technicians with certificate, and one with diploma, in metalworking.

Firm 581. One engineer trained abroad, one accountant and one MBA locally trained. Six technicians had metalworking certificates.

Firm 1141. No professionals. One technician with diploma in data processing and one in management.

Firm 1641. One engineer trained abroad and one accountant trained locally. Two technicians with certificate and three with diploma in metalworking-engineering.

Firm 411. Among professional personnel: two chemists, nine accountants, one MBA, and one bachelor with a major in business studies, all locally trained. Among technical personnel: five electricians and one electronics technician with certificate, and two technicians with diploma in data processing.

Firm 491. Professionals included two engineers trained locally and one abroad, plus two university graduates in personnel and production. Among technical personnel, one with diploma in metalworking and one with diploma in food technology, plus 26 with certificate in metalworking, one in woodworking, nine in printing technology and one electrician. Two of the metalworking technicians were expatriates.

Firm 1361. No professionals. Among technical personnel, six technicians with diploma in metalworking.

Firm 1751. One engineer (Chief Executive) trained abroad. Among middle-level and technical personnel there were two with certificate in data processing, plus five staff with diplomas in: finance, accounting and administration, metalworking, personnel and sales and marketing. Professional and technical personnel in metalworking are shown in Table AIV-8.

A comparison of skilled personnel for all four industries surveyed is shown in Table AIV-9.

The comparison shows that, as hypothesized, food and beverages and metalworking require, due to technological reasons, a higher proportion of professional personnel. Such personnel is practically lacking in wood and furniture.

Table AIV-8 Metalworking. Professional and technical personnel

Firm	*Profess.*	*Diploma*	*Certifi.*	*Employ.*	*% Profes.*	*% D + C*
521	1	1	3	24	4.2	16.7
581	3	0	6	150	2.0	4.0
1141	0	2	0	88	0.0	1.1
1641	2	3	2	96	2.1	5.2
411	13	2	6	1353	1.0	0.6
491	5	2	37	528	0.9	7.4
1361	0	6	0	43	0.0	13.9
1751	1	5	2	75	1.3	9.3
Total	**25**	**21**	**56**	**2357**	**1.1**	**3.3**

Notes:
D: Diploma
C: Certificate

On the other hand, a relatively high proportion of middle-level technical personnel with certificates and diplomas was found in wood and furniture. The total, average, figures on this item were however distorted by the weight of one food and beverages firm's data.

Table AIV-9 Professional and technical personnel, all industries

Industry	*% Professionals*	*% Certif. + Diploma*
Food and Beverages	1.10	7.03
Textiles and Clothing	0.80	1.30
Wood and Furniture	0.30	3.90
Metalworking	1.10	3.30
Average	**0.90**	**6.90**

Notes and References

1. The points of view expressed in this work are those of the author and do not purport to represent the official position of any international organization, their affiliates or directors.
2. These surveys were sponsored by The World Bank. For the study carried out in Ghana, see Lall, Navaretti, Teitel and Wignaraja, *Technology and Enterprise Development. Ghana Under Structural Adjustment*, London: Macmillan, 1994. A similar study on Kenya is in preparation.
3. For the history of Zimbabwe settlers until independence, see Mosley (1983) and Phimister (1988).
4. The Shonas are divided in six groups, of which the largest, Karanga, lives in the far west of the country, separated from the other Shona groups by the Ndebeles. Politically, the Karangas generally lean in the direction of the Ndebele.
5. An experience not unlike that of some Latin American industrializing countries that developed export-oriented industries from import substitution activities originally started under protection. See Teitel and Thoumi (1986).
6. For an examination of politics and economics in Zimbabwe before the adjustment programme, see Stoneman and Cliffe (1989), and Baynham (1992).
7. For an examination of the land reform record of the decade 1980–90, see Bratton (1990).
8. It has been estimated that the white population, that was around 270,000 before independence, has now dwindled to less than 80,000.
9. This mentality may also have been unwillingly fostered by some international aid donors.
10. For an examination of protection of industry in Zimbabwe, see Ndlovu (1994).
11. In 1979, Zimbabwe's violent death rate was second in the world only to El Salvador's in Central America.
12. Scandals have been rumoured in connection with telephones, construction of the new airport, the acquisition of planes for Air Zimbabwe, and ZISCO, the government-owned iron and steel corporation ('Palm oil greases the wheels: Zimbabwe', *The Economist*, March 2, 1996, p. 44).
13. It is claimed that the country produces tractors with up to 70% local content and about 90% of all diesel engines are manufactured locally. There is also local production of fertilizers but using a high proportion of imported inputs. The extent to which industry could be based on agricultural development is discussed in Mair (1992).
14. The production volume growth rates are derived from Nelson (1983, appendix); however, there are no manufacturing value added growth rate data for this period.

15. These high growth rates probably resulted from the need to substitute imports of spare parts, simple metal consumer products, and fabrics, and the growth of domestic apparel demand *pari passu* with the growth of population.
16. For the apparent reorientation of the economic structure toward primary goods, as well as other consequences of the liberalization programme for growth and distribution, see Davies and Rattso (1996).
17. Another, ominous reason for recent declines in population growth is the high incidence of AIDS-related mortality.
18. For the theoretical rationale, see Salter (1960, chapter IV).
19. In general, foreign firms, and large local plants, tend to hire a higher proportion of black managers than small local enterprises.
20. One of the few bright spots in manufacturing during 1996 was the high demand for locally assembled automobiles (mostly Japanese models) which reached about 900 vehicles per month (*EIU Country Report, 4th quarter, 1996*).
21. See Teitel (1994a) and (1994c) for studies relating such science and technology output indicators to country size and per-capita income. Teitel (1987) applies similar methods to relate science and technology *inputs* to the same variables.
22. For example, in 1980, 10 Latin American countries had each more than 100 patents granted with three of them having each more than 1,000. For these countries, the average (non-weighted) proportion of patents granted to residents was 11% (Inter-American Development Bank, 1988, table H-4).
23. Of course, industrial R&D may also be carried out outside the firm, at government or industry laboratories or research institutes.
24. This larger sample was originally obtained by stratified cluster sampling based on enterprise size determined by employment, Economic and Social Institute, Free University, 'First Report on the Zimbabwe Survey', Amsterdam, 1993.
25. To compare with Ghana, see Lall *et al.* (1994).
26. For more details, see Teitel (1981b), and (1993, chapter 11).
27. Hirschman attributed to machine-paced processes the need for lower operative skills and thus argued that, paradoxically, more mechanized (and perhaps also more capital-intensive) manufacturing processes could lead to a smaller productivity gap between industrialized and developing countries. See Hirschman (1958, p. 152) for the proposed hypothesis, and Teitel (1981a) for testing of the same.
28. Similar restrictive business practices have been opportunely reported also in Latin America; see Vaitsos (1975).
29. Although electricity rates were not higher at night, higher rates did apply during peak demand periods. One-shift operations can avoid the peak period but a two-shifts operation cannot.
30. In one plant, management claimed to have in operation a statistical quality control system which is a sophisticated quantitative method of quality control. However, since there was no evidence of statistical control charts or records at the production site, this answer was disallowed.

31. For studies on the application of Japanese management techniques in Zimbabwe, see Kaplinsky (1995) and Posthuma (1995).
32. Although the recipient firms were in the private sector and not always small.
33. One manager proudly pointed to a young man who had advanced from floor sweeper to computer operator in only two years.
34. However, one plant in this sector is already implementing a massive renovation and reconstruction programme that also has substantially affected its organization.
35. While as shown in formula (1) the three partial numerical indicators of technology transfer, skills and technical effort were simply added up to construct the technology variable, alternative formulations, e.g. multiplicative, logarithmic, etc., could be postulated.
36. See, for example, Stiglitz (1986, ch. 4) and Stokey and Zeckhauser (1978, ch. 14).
37. Steel and petrochemical projects undertaken in the 1970s in Brazil represent good examples.
38. See Rodrik (1996) for a good discussion of the need to incorporate such mechanisms when undertaking economic policy reforms in developing countries.
39. For a discussion of the influence of manufacturing exports fostering the adoption of more advanced technologies and productivity improvements in East Asian countries, see World Bank (1993c, chapter 6).
40. See Lall (1996, ch. 3) for a good description of technology development policies recently implemented in several Asian countries.
41. The WTO-UNCTAD International Trade Centre (ITC) provides export-oriented technical assistance, and the International Labour Office (ILO) has set up productivity centres, or programmes, in a number of industrializing countries. The assistance provided by these institutions could be beneficially used in Zimbabwe, with account taken that close coordination with the private sector is required.
42. Financiadora de Estudios y Proyectos (FINEP) in Brazil, is a government agency reputed to have had successful experience in this area. Its own sources of funds have been complemented by science and technology loans from the Inter-American Development Bank and the World Bank.

Bibliography

Baynham, S. (ed.) (1992) *Zimbabwe in Transition*, Stockholm: Almquist & Wiksell International.

Bratton, M. (1990) 'Ten Years After: Land Redistribution in Zimbabwe, 1980–1990', in R. L. Prosterman, M. Temple and T. M. Hanstad (eds), *Agrarian Reform and Grassroots Development: Ten Case Studies*, Boulder and London: Lynne Rienner, pp. 265–91.

Cook, P. and F. Nixson (eds) (1995) *The Move to the Market? Trade and Industry Policy Reform in Transitional Economies*, New York and London: St. Martin's Press and Macmillan Press.

Davies, R. and J. Rattso (1996) 'Growth, Distribution and Environment: Macroeconomic Issues in Zimbabwe', *World Development*, (February), vol. 24, no. 2, pp. 395–406.

EIU *Zimbabwe. Country Report*, London: The Economist Intelligence Unit, 1993, 1994, 1995, 1996 and 1998 issues.

Fallon, P. R. and R. E. B. Lucas (1993) 'Job Security Regulations and the Dynamic Demand for Industrial Labor in India and Zimbabwe', *Journal of Development Economics*, 40(2), (April), pp. 241–75.

Hirschman, A. O. (1958) *The Strategy of Economic Development*, New Haven: Yale University Press.

Inter-American Development Bank (1988) *Economic and Social Progress in Latin America, 1988 Report*, Washington, DC.

Kaplan, I. (1983) 'The Society and Its Environment', in H. D. Nelson (ed.), *Zimbabwe: A Country Study*, Area Handbook Series, United States Government as Represented by the Secretary of the Army.

Kaplinsky, R. (1995) 'Technique and System: The Spread of Japanese Management Techniques to Developing Countries', *World Development*, 23(1), (January), pp. 57–71.

Lall, S. (1996) 'Learning from the Asian Tigers', *Studies in Technology and Industrial Policy*, London: Macmillan.

Lall, S., G. B. Navaretti, S. Teitel and G. Wignaraja (1994) *Technology and Enterprise Development: Ghana Under Structural Adjustment*, London and New York: Macmillan and St. Martin's Press.

Mair, S. (1992) 'Agricultural Demand-Led Industrialization – An Option for Zimbabwe?', in H. H. Bass *et al.* (eds), *Industrialization Based on Agricultural Development*, African Development Perspectives Yearbook 1990/91, Vol. 2, Munster: Lit., pp. 556–76.

Mead, D. C. (1994) 'The Contribution of Small Enterprises to Employment Growth in Southern and Eastern Africa', *World Development*, 22(12), (December), pp. 1881–94.

Mosley, P. (1983) *The Settler Economies: Studies in the Economic History of Kenya and Southern Rhodesia, 1900–1963*, New York: Cambridge University Press.

Ndlovu, L. B. (1994) *The System of Protection and Industrial Development in Zimbabwe*, Aldershot, England: Ashgate Pub. Co.

Nelson, H. (ed.) (1983) *Zimbabwe: A Country Study*, Area Handbook Series, United States Government as Represented by the Secretary of the Army.

Nelson, R. R. (1992) 'What is Commercial and What is "Public" About Technology, and What Should Be?', in N. Rosenberg, R. Landau and D. C. Mowery (eds), *Technology and the Wealth of Nations*, Stanford: Stanford University Press, pp. 57–71.

— (ed.) (1993) *National Innovation Systems. A Comparative Analysis*, New York and Oxford: Oxford University Press.

Phimister, I. (1988) *An Economic and Social History of Zimbabwe, 1890–1948: Capital Accumulation and Class Struggle*, London and New York: Longman.

Posthuma, A. C. (1995) 'Japanese Techniques in Africa? Human Resources and Industrial Restructuring in Zimbabwe', *World Development*, 23(1) (January), pp. 103–16.

Riddell, R. C. (1988) *Industrialization in Sub-Saharan Africa: Country Case Study – Zimbabwe*, ODI Working Paper 25, London: ODI.

— (1990) *Manufacturing Africa. Performance and Prospects of Seven Countries in Sub-Saharan Africa*, London: ODI with Heinemann and Currey, ch. 10.

Rodrik, D. (1996) 'Understanding Economic Policy Reform', *Journal of Economic Literature*, (March), vol. XXXIV, no. 1, pp. 9–41.

Salter, W. E. G. (1960) *Productivity and Technical Change*, 2nd edn, Cambridge: Cambridge University Press.

Skalnes, T. (1993) *The State, Interest Groups and Structural Adjustment in Zimbabwe*, Ilford Essex: Frank Cass.

Stiglitz, J. E. (1986) *Economics of the Public Sector*, New York: W. W. Norton & Co.

Stokey, E. and Zeckhauser, R. (1978) *A Primer of Policy Analysis*, New York: W. W. Norton & Co.

Stoneman, C. (1992) 'Policy Reform or Industrialization? The Choice in Zimbabwe', in R. Adhikari, C. Kirkpatrick and J. Weiss (eds), *Industrial and Trade Policy Reform in Developing Countries*, New York: St. Martin's Press, pp. 97–110.

Stoneman C. and L. Cliffe (1989) *Zimbabwe: Politics, Economics and Society*, London and New York: Pinter Publishers.

Teitel, S. (1981a) 'Productivity, Mechanization and Skills: A Test of the Hirschman Hypothesis for Latin American Industry', *World Development*, vol. 9, no. 4, pp. 355–71.

— (1981b) 'Towards an Understanding of Technical Change in Semi-Industrialized Countries', *Research Policy*, vol. 10, pp. 127–47.

— (1984) 'Technology Creation in Semi-Industrial Economies', *Journal of Development Economics*, vol. 16, nos 1–2, pp. 39–61.

— (1987) 'Science and Technology Indicators, Country Size and Economic Development: An International Comparison', *World Development*, vol. 15, no. 9, pp. 1225–35.

— (1993) *Industrial and Technological Development*, Washington DC: Inter-American Development Bank/The Johns Hopkins University Press.

— (1994a) 'Patents, R&D Expenditures, Country Size, and Per-Capita Income: An International Comparison', *Scientometrics*, vol. 29, no. 1, pp. 137–59.

— (1994b) *Technology and Skills in Kenyan Manufacturing*, Washington DC: Africa Technical Department, The World Bank.

— (1994c) 'Scientific Publications, R&D Expenditures, Country Size, and Per-Capita Income: A Cross-Section Analysis', *Technological Forecasting and Social Change*, vol. 46, no. 2, pp. 175–187.

Teitel, S. and F. E. Thoumi (1986) 'From Import Substitution to Exports: The Manufacturing Exports Experience of Argentina and Brazil', *Economic Development and Cultural Change*, vol. 34, no. 3, pp. 455–90.

The Economist (1996) 'Palm Oil Greases the Wheels: Zimbabwe', *The Economist*, March 2, 1996, p. 44.

Thomas D. and I. Murandi (1994) 'The Demographic Transition in Southern Africa: Another Look at the Evidence From Botswana and Zimbabwe', *Demography*, 31(2), (May), pp. 185–207.

UNESCO (1993) *Statistical Yearbook*, Paris.

Vaitsos, C. (1975) 'The Process of Commercialization of Technology in the Andean Pact', in H. Radice (ed.), *International Firms and Modern Imperialism*, Harmondsworth: Penguin Books, pp. 183–214.

Watanabe, S. (1987) 'Technical Capability and Industrialization. Effects of Aid and Sanctions in the United Republic of Tanzania and Zimbabwe', *International Labor Review*, 128(5), (Sept.–Oct.), pp. 525–41.

Westphal, L. E. (1982) 'Fostering Technological Mastery by Means of Selective Infant-Industry Protection', in Syrquin and Teitel (eds), *Trade, Stability, Technology and Equity in Latin America*, New York: Academic Press, pp. 255–79.

Whiteside, A. W. (ed.) (1989) *Industrialization and Investment Incentives in Southern Africa*, Portsmouth, New Hampshire: Heinemann Educational Books.

World Bank (1993a) *World Development Report 1993*, published for The World Bank. New York: Oxford University Press.

World Bank (1993b) *Zimbabwe Country Background Paper*, Economic and Social Institute, Free University, Amsterdam, and Department of Economics, University of Zimbabwe, Harare, April 1993, The World Bank.

World Bank (1993c) *The East Asian Miracle: Economic Growth and Public Policy*, Published for the World Bank, New York: Oxford University Press.

World Intellectual Property Organization (1983) *100 Years Protection of Industrial Property – Statistics – 1883–1982*, Geneva.

— (1986) *Industrial Property Statistics, 1985*, Geneva.

World Resources Institute (1993) *World Resources, 1992–93*, New York: Oxford University Press.

Zymelman, M. (1990) 'Science, Education, and Development in Sub-Saharan Africa', Technical Paper No. 124, Africa Technical Department Series, Washington DC: The World Bank.

Index